D0983534

Newnes
Circuit Ideas
Pocket Book

Newnes

Circuit Ideas
Pocket Book

Newnes
An imprint of Butterworth-Heinemann Ltd
Linacre House, Jordan Hill, Oxford OX2 8DP

Ⓡ A member of the Reed Elsevier plc group

OXFORD LONDON BOSTON
MUNICH NEW DELHI SINGAPORE SYDNEY
TOKYO TORONTO WELLINGTON

© Reed Business Publishing

All rights reserved. No part of this publication
may be reproduced in any material form (including
photocopying or storing in any medium by electronic
means and whether or not transiently or incidentally
to some other use of this publication) without the
written permission of the copyright holder except
in accordance with the provisions of the Copyright,
Designs and Patents Act 1988 or under the terms of a
licence issued by the Copyright Licensing Agency Ltd,
90 Tottenham Court Road, London, England W1P 9HE.
Applications for the copyright holder's written permission
to reproduce any part of this publication should be addressed
to the publishers

British Library Cataloguing in Publication Data
A catalogue record for this book is available from the
British Library

ISBN 0 7506 2336 5

Library of Congress Cataloguing in Publication Data
A catalogue record for this book is available from the
Library of Congress

Printed in England by Clays Ltd, St Ives plc

Contents

Automatic generator for VGA sync

Low-priced colour computer monitors often seen on the surplus market accept negative-going line sync, which makes them incompatible with the VGA standard used by many PCs. This circuit produces negative-going sync. whatever the polarity of the input.

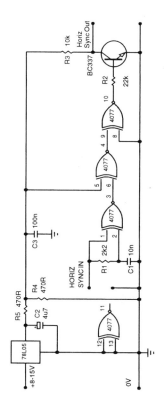

Automatic sync plunger produces negative sync. regardless of the input polarity.

Dividing the 5V line in the R_4, R_5 potential divider to 2.5V eliminates the need to amplify the input signal. A CR network R_1C_1 on the input of an exclusive-Nor holds pin 2 high when a negative sync is applied, the pulses being too narrow for C_1 to discharge below the gate threshold; the syncs are therefore not inverted. If the input is positive-going, pin 2 is not allowed to rise and the input is inverted.

A buffer, an inverter and a *BC337* current amplifier prepare the signal for combining with the vertical syncs.

Tom Sheppard

Swaythling
Southampton

Programmable timer

Three seconds to nearly six hours is the range of this crystal-controlled timer, at a resolution of 1s. In comparison with D Ibrahim's design in September 1990 *Circuit Ideas*, the time range is increased by a factor of 200, accuracy is better and fewer ICs are needed.

Four thumb-wheel switches set the time interval. Initially, the counter IC_2 is inhibited and preset,

Programmable timer covers the 3s-6h range at 1s resolution. Crystal frequency generator is shown, but other means would suffice.

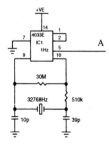

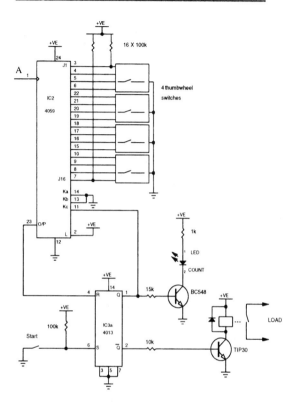

inputs to Ka,b,c being low. A start signal from the switch sets the D-type flip-flop IC_3 and the counter begins to count, the Q output of IC_2 resetting the flip-flop at time-out. During the count cycle, the led illuminates and the relay is on.

The *CD4013* generates 1Hz pulses at crystal accuracy, but any other type of pulse generator would be suitable.

V B Oleinik

Kaliningrad Moscow Region Russia

LC square-wave generator

As an alternative to the usual RC arrangement, an LC circuit offers better frequency stability. This Colpitts oscillator circuit provides a square wave, using an op-amp as the active element.

Any of the standard op-amps will work reasonably well at audio frequencies, but an op-amp such as an *LM6361* would be better at higher frequencies.

Output frequency is $1/(2\pi\sqrt{(LC/2)})$, the two capacitors being effectively in series.

Michael A Covington

Athens
Georgia
USA

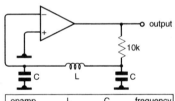

opamp	L	C	frequency
Any	33 mH	0.04 μF	6 kHz
LM6361	470 μH	40 pF	1.3 MHz

LC square-wave generator circuit shown with component values required for two typical frequencies and op-amps.

Fast, full-wave rectifier

Precision rectifiers using op-amps with feedback diodes perform extremely well, except where speed is concerned, bandwidth and slew rate being limiting factors. This circuit overcomes the speed barrier.

Transistor Q_1 is a common-emitter amplifier and Q_2 is connected in common-base configuration, so that each half-cycle of input draws current through R_7. The result is a full-wave rectifier. $R_{1,2,3}$ and $R_{5,6}$ define emitter currents and $C_{1,2}$ speed up the action.

With R_7 at 1.8kΩ, output is 50% input, the -3dB point occurring beyond 2MHz. If a milliammeter is used in parallel with a 560Ω resistor (R_{7a}) and a 100nF smoother, a ±5V sine input gives 50% FSD meter deflection with a response past 20MHz. Non-linearity at 100kHz is less than 5% FSD on the meter.

C J D Catto

Elsworth
Cambridgeshire

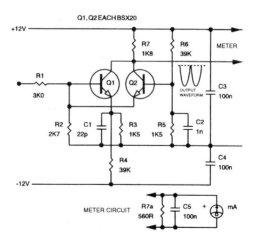

Precision rectifier compares in accuracy with the op-amp/diode variety and exhibits a 2MHz bandwidth.

Low-cost digital phasemeter

To make a phase meter with a typical error of 0.2%, the ICL 7107 digital voltmeter IC can be pressed into service to display the phase difference of two signals in degrees.

Two LM311 comparators convert the inputs to cmos-compatible square waves. The leading edge of the square at IC_1 output sets FF_1 and removes the inhibit from FF_2, that from U_2 setting FF_2 which resets FF_1, FF_2 in turn being reset. Output at Q_1 is therefore a pulse whose width is directly proportional to phase difference between the inputs.

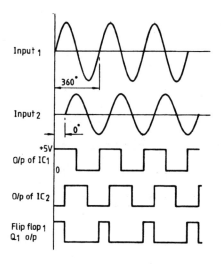

Resistors R_5 and R_6 form a voltage divider to provide a voltage reference of 1.02V for the ICL7107, the height of the pulses from Q_1 being set by R_3 and R_4 to $3.6V_{ref}$ at position 1 of the range switch. The average voltage of the pulse train is $(3.6V_{ref} \times$ duty cycle$)$ and the display reads $(1000 \times$ average$/v_{ref})$, which is $3600 \times$ phase difference$/360°$, or $10 \times$ phase difference.

The decimal point of the display is arranged to give a reading in degrees, so that the resolution is $0.1°$ over the range 0-199.9. For the range set by resistors $R_{1,2}$, measurement resolution is $1°$ over 0-360°. Measurement accuracy is better than 0.2% over 1kHz to 250kHz, falling off at lower frequencies.

M.S. Nagaraj

ISRO Satellite Centre
Bangalore India

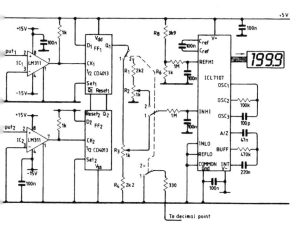

To decimal point

Battery life extender

Since primary batteries are usually discarded when their output voltage has dropped below about two thirds of the initial voltage, any means of reducing the consequent waste of money and resources is worth considering. The problem is to ensure that the performance of battery-powered equipment does not suffer at a reduced battery voltage.

In this circuit, a TLC3702 comparator senses when the battery output has reached a level determined by R_1/R_2; resistor R_3 provides a degree of positive feedback to avoid oscillation. At this point, the three-gate oscillator driving a 74HC4053 capacitive multiplier increases the output to 150% of the battery voltage, which enables the battery to continue in service until its voltage has dropped to less than 50%.

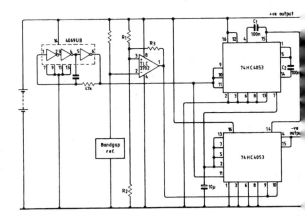

The multiplier is similar to the 7660 capacitive doubler but, instead of charging a capacitor and placing it series with the battery, Two capacitors (C_1 and C_2 are charged in series and placed in series with the battery but in parallel with each other. Although the oscillator shown runs at around 100kHz, frequency is not important, C_1 and C_2 being adjusted to suit.

Efficiency of the circuit is about 90% at 2mA, falling to 80% at 5mA. Several switch packages can be used in parallel to improve efficiency.

Ian M. Wiles

IPR Technology

Basingstoke

Hampshire

"Fuse-blown" indicator

When the fuse blows, Led1 indicates the fact. Normally, the red led, D_1, is in parallel with the green led D_2, the voltage drop across D_1 preventing Led_1 from illuminating. A blown fuse removes the current through Led_2, the voltage across Led_1 allowing it to light. Resistor $R=DC_{in}/0.01$, unless DC_{in} is 5V, in which case $R=470\Omega$.

T. Cottignoli

Taranto,

Italy

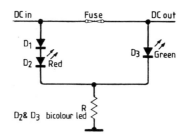

Composite-feedback amplifier

An amplifier with a defined output impedance can be implemented rather more easily than in the circuit given by A.J. Chamberlain in the October 1989 Circuit Ideas by the use of a modified version of the Howland current-pump circuit shown left.

Voltage feedback to inverting and non-inverting inputs cancels, leaving only current feedback and, therefore, a high output impedance. To produce a specific output impedance, reduce the amount of feedback to the non-inverting amplifier input by the appropriate amount. For example, changing R_{2b} to $22\text{k}\Omega$ gives a Z_{out} of 300Ω.

Two current pumps can be driven in antiphase, a resistor being connected between the non-inverting inputs to produce an amplifier with a floating output giving +24dB with 24V rails and +16dB with 6V rails.

D. Austin
Birmingham

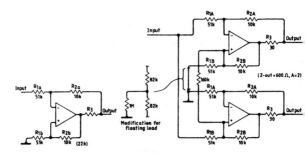

Precision pulse-width generator

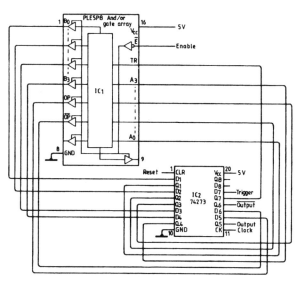

Programmable pulse generators commonly have limited programming capability and exhibit initial timing error caused by lack of synchronism between the input trigger and the system clock. The circuit described overcomes these disadvantages. No RC timing is used and accuracy is solely dependent on clock frequency; a wide range of pulse widths is obtained by changing the clock frequency.

A PLE5P8 programmable logic element by Monolithic Memories and a 74273 octal latch compose the circuit; the clock input to the 74273 may be the system clock. Four of the five inputs to the PLE are used for state-incrementing control, the fifth serving as the trigger. The clear input of the latch functions as the reset input for the generator. Both active-high and active-low outputs are available.

When the trigger goes low, the true output of the generator goes high and begins to time-out the programmed n clock cycles; after one complete cycle, the true output goes low. In the case of re-triggering, timing continues for another n cycles and, if re-triggering continues, so does the output timing.

In the circuit shown, from 1 to 16 clock cycles can be programmed at any desired frequency, and a PLD

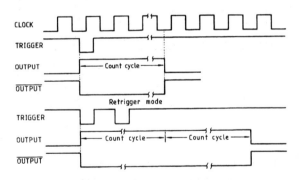

with more inputs will allow a greater selection of timing combinations; a PLD with nine inputs will give 1 to 256 clock cycles, one input being used as the trigger input.

V.Lakshminarayanan

*Centre for Development of Telematics
Bangalore
India*

High frequency switch

An emitter follower can be used as an RF switch, which will work at VHF, with a high switching speed.

Transistor Tr_1 is an emitter follower and Tr_2 a current switch. A voltage of 0V at the control input cuts off Tr_2 and therefore D_1, allowing Tr_1 to function as an emitter follower. When the control voltage is 12V, Tr_2 is on, D_1 conducts and cuts Tr_1 off, preventing signal reaching the output. Attenuation in this condition is greater than -30dB at 80MHz. The control input is cmos-compatible, but the circuit shown right can be used to make the circuit usable for TTL input.

D.I Malynovsky

Leningrad
USSR

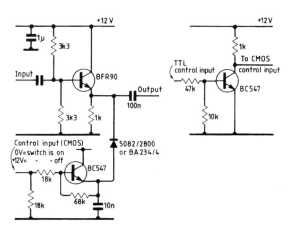

Absolute-value differencer

Using a single battery supply, the outputs of these circuits are (V_a-V_b) or (V_b-V_a), whichever gives a positive answer, which is the magnitude or absolute value of (V_a-V_b).

In the simple version (top), the more positive input, selected by D_1 and D_4 — a "highest wins" gate — to the non-inverting input. Diodes D_2 and D_3, which form a "lowest wins" gate, select the smallest input and feed it to the inverting input. The output is therefore the larger of V_a or V_b minus the smaller of V_a or V_b.

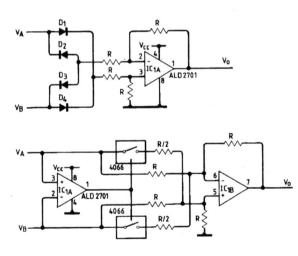

One disadvantage of the circuit is the forward voltage drop of the diodes. An improvement is the lower circuit, in which the diode gates are replaced by analogue switches controlled by an ALD2701 op-amp comparator, which has a good common-mode input voltage range and output swing and is therefore very suitable for battery-powered circuits. If a fast op-amp or a dedicated comparator IC were used in this position, a degree of hysteresis would be needed to avoid oscillation at the switching point.

M. Neal

London SW15

On-board transistor tester

This instrument will test transistors without removing them from the printed board. A 555 multivibrator oscillates at a frequency of 1kHz, its output being taken to a 7474 D-type flip-flop connected as a toggle, which produces complementary square waves (Q and /Q) at 500Hz. Diodes D_5 and D_6 are red and green leds in the same package. Base drive for the transistor under test comes from the mid-point of the Q and /Q potential divider.

With no transistor connected, the bicolour led appears amber, since both leds switch on and off at 500kHz. If a good p-n-p transistor is connected, it is on when Q is low and /Q is high, since its base/emitter is forward-biased; in this condition, neither led lights since a low Q reverse-biases D_5 and the voltage across D_6 is equal to $V_{CE(on)}$ which, for a good transistor is 0.1V. During the next pulse, Q becomes high and /Q low and a good transistor at the terminals will be off; in this condition, D_6 is off because it is reverse-biased and D_5 is on. The opposite effect applies if a good n-p-n transistor is connected.

A good transistor has a collector/-emitter voltage of around 0.1V and a silicon diode drops about 0.6V. Each of the two loops formed by $D_{1,2}$ and $D_{3,4}$ between pins 5 and 6 of the flip-flop has one collector/emitter drop and two diode drops to impress on the leds which, at $0.1+1.2 = 1.3V$, is insufficient to turn on one of the leds which stays off if the transistor is good and on if the device has one shorted junction to make it behave like a diode; in this case the voltage drop is $1.2 + 0.6 = 1.8V$.

Therefore, one led lights if the transistor is good (both p-n-p and n-p-n); both leds are off for a transistor with shorted C/E junction; and both are on if the transistor is bad. Diodes $D_{1,2,3,4}$ prevent false indications of normality when a transistor with a B/C short or a B/E short is connected.

V. Lakshminarayanan

Centre for Development of Telematics
Bangalore India

state	D_5	D_6
open C/E	flicker	flicker
short C/E	off	off
good p-n-p	on	off
good n-p-n	off	on

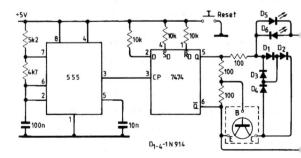

Courtesy-light delay

This circuit provides a car courtesy-light delay. It holds the interior light on for a few seconds to allow seat belts to be fastened and the ignition key inserted but, when the battery voltage falls as the engine turns over, the light goes out.

In the quiescent state, C_1 is charged to almost 12V, C_2 has both ends at about 12V and both transistors are off. When the door is opened, both ends of C_2 are taken to ground, base current flows through R_1, C_1 discharges and both transistors and the lamp come on.

Closing the door opens the switch and C_2 begins to charge through the base of Tr_1. When C_2 charges to the point where Tr_1 cannot hold Tr_2 on, the action becomes cumulative, since the bottom end of C_2 starts to go positive, switching Tr_1 off.

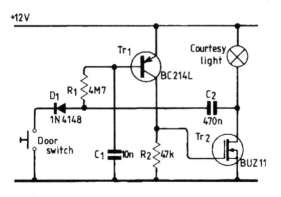

When the starter motor turns, voltage across the circuit drops slightly and, since the voltage across C_1 cannot change quickly, the emitter of Tr_1 is taken more negative than its base and it turns off in another cumulative action.

Resistor R_2 ensures that leakage in Tr_1 does not turn Tr_2 on accidentally and D_1 isolates other lamps operated by the switch, which would otherwise cause a very short delay.

Chris Miller

Sevenoaks

Economical switch debouncer

A very low-cost push-button switch debouncer can be made using a couple of spare tri-state buffers used as inverters.

In the diagram, when the enable line of the buffers is high, the buffers are in the high-impedance state, their outputs being pulled low by the 470ohm resistors; when enable is low, the buffer outputs are pulled high by the 2.2kohm resistors. The effect is that the enable line is used as input for the inverting operation. Cross-coupling between the buffers forms a latch, which is used to debounce the switch.

V. Lakshminarayanan

Centre for Development of Telematics
Bangalore India

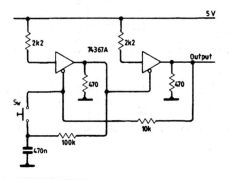

High-frequency digital oscillator

Needing a high-frequency digital crystal oscillator for use in a frequency synthesiser, I decided to save a few components by using a comparator as the oscillator to produce a square wave directly, instead of using a sine-wave oscillator and squaring the output with the comparator.

The circuit is similar to other oscllators of this type using one inverting gate, but not many gates can match the speed of the AD9685, which I used to produce ECL levels; for TTL levels, I would suggest the AD9686. A 10MHz waveform is obtained reliably, but I see no reason why much higher frequencies should not be obtained, perhaps with component value changes. Pay particular attention to the loading of the crystal. Using the 9685 gives the facility to drive a 50Ω load if the output is properly terminated.

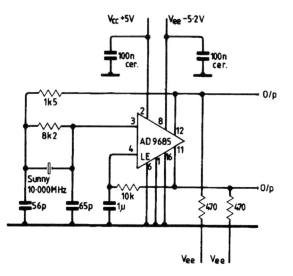

Since the 9685 will oscillate at over 500MHz, a large, low-inductance ground plane is needed, as are short lead lengths and decoupling capacitors close to the supply pins. Analog Devices advise against the use of IC sockets.

Phil Denniss

Department of Plasma Physics
University of Sydney
Australia

Single-phase to three-phase converter

This is a method of deriving a three-phase sine-wave reference using a frequency tracking phase-shift network.

In the diagram, the analogue multiplier IC_1, with R and C, forms a voltage-controlled transfer-function generator, of which the transfer function is

$$(V_o/V_{in})(s) = (V_{CM}RCs)/(10+V_{CM}RCs) \quad (1)$$

where V_{CM} is the multiplier control voltage.

The input signal V_{IN} is half-wave rectified by IC_2

$V_B = |V_{in}| \angle 120 \text{deg}$

V_0

60deg

$V_A = V_{in}$

$V_C = |V_{in}| \angle -120 \text{deg}$

and the multiplier output V_O is half-wave rectified by IC_3 after being amplified by a factor of 2 in IC_3. Integrator IC_4 forms a low-pass filter and high-gain comparator for these two rectified signals, its output being the control voltage V_{CN} of the multiplier. This feedback loop maintains the magnitude of the transfer function at 0.5 and, under this condition, the phase of the transfer function is exactly 60°.

The phasor diagram shows how the other two phases are derived from the output of the transfer-function generator. If $V_A = V_{IN}$ is the reference phase, V_O has half the amplitude of V_A and leads it by 60°. In conjunction with the transfer-function generator, IC_6 forms an all-pass network which doubles the phase lead of V_O while restoring the amplitude to that of V_A to form the second phase V_B. The third phase V_C is formed by inverting V_O and amplifying it by 2 in instrumentation amplifier IC_5. Input to the circuit must be sinusoidal with no DC offset, although input amplitude is not important. Operation is from 5Hz to 100Hz, distortion is less

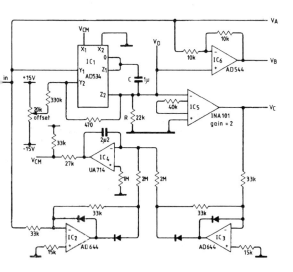

than 2%, phase error is better than 0.75degrees and the input should be between 1V and 5V.

Ajoy Raman and K. Radhakrishna Rao

Indian Institute of Technology
Madras

40W voltage doubler

A cheap audio power IC, the TDA 2004/5, will serve to make a regulated voltage doubler giving up to 4A at 20V from a 12V car battery. The original circuit was designed to power a car radio amplifier, which needs an 18V, 4A(pk) regulated supply.

With a few extra components, the IC is used as a switching H bridge that includes thermal and overvoltage shutdown and current limiting. Since the transition times are 200ns when the IC is driven by

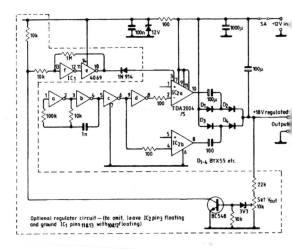

more than 10V, switching rates of up to 300kHz can be used. In this circuit, the frequency is 35kHz; standard 85°C electrolytics give 2A continuous in an ambient temperature of 50°C or less. For higher temperatures, use 105°C, low-ESR capacitors. Efficiency is 80% for loads of 500mA to 1.5A; Schottky diodes could be used for higher efficiencies, but are expensive. The output voltage stays within 0.1V from no load to 2A and ripple is about 0.1Vpk-pk; no-load current is only 10mA. Thermal shutdown occurs after a few seconds when the output is loaded to 4A on a 13.5V supply.

Ian Hegglun

Manawatu Polytechnic
New Zealand

RC attenuator distortion

Figure 1(a) is a typical 10:1, 1MΩ wide-band attenuator, often used in signal generators, millivoltmeters and oscilloscopes. It is compensated by C_1 and C_2, so that $R_1/R_2 = C_1/C_2$ (neglecting C_{pcb}), and attenuation ought to be constant for all input frequencies, depending on source impedance and input capacitance. Unfortunately, one cannot neglect C_{pcb}, particularly since it is not constant with frequency and cannot therefore be cancelled by adjustment of C_1. Special PCB materials can be used which do have constant electrical properties, but they are expensive.

  A step function passed through the attenuator exhibits the effect seen in Figure 2, which shows what happens with adjustment of C_1; the "hook" is

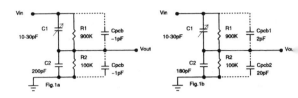

Figure 1. At (a), a typical 1MΩ, 10:1 RC attenuator, showing PCB strays, which are not constant with frequency and introduce a "hook" in a step function. Circuit at (b) is a complete cure; artificial "strays" in proportion to attenuation introduce impedance changes in each branch that compensate each other. Trim the two additional Cs and then adjust C_1 for an ideal response.

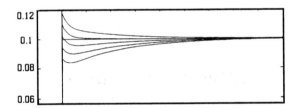

Figure 2. Without the two additional pads, this is the attenuator response to a step function. Whatever the setting of C_1, the hook in the response stays due to dielectric adsorption.

ever-present, regardless of C_1 setting and makes its presence felt mainly in the 10-200kHz band with the values shown in Figure 1. Its amplitude is roughly $C_{pcb}/(C_{pcb} + C_1)$.

Using a ground plane around the output is not totally effective, since C_2 now has a great deal more capacitance to contend with. Instead, my solution is to make a pair of "deliberate strays", C_{pcb1} and C_{pcb2}

in Figure 1(b), using pads on both sides of the board with areas in proportion to the desired attenuation. Trimming the pads to exact size by drilling small holes allows complete cancellation of the hook. Figure 3 gives a suggested layout.

Erik Margan

Ljubljana
Slovenia

GND Vout

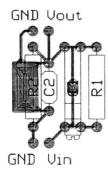

Figure 3. Suggested board layout of Fig.1(b) circuit. The track area is in the ratio of C_1 to C_2.

GND Vin

1:1 square waves with 2ns edges

With a bit of care in layout, this square-wave generator will produce accurate 5V output of 50% duty cycle at 1MHz or 1kHz with transition times of less than 2ns.

It is composed of an emitter-coupled Schmitt-trigger oscillator, $T_{1,2,3}$, its RC feedback components being R and C_1 or C_2, switched for 1MHz or 1kHz. The fast rise time and accurate level control are the responsibility of the two UHF transistors, which form a current-mode switch. Low-tolerance, metal-film resistors and adjustable IC regulators for the 5V and -6.9V supplies supply the accuracy to within 1%.

To calibrate the oscillator, first adjust the 5V. Connecting T_2 base to negative by 100Ω stops oscillation and turns T_5 on; adjust the -6.9V toobtain 0V at the output. Disconnect the 100Ω resistor and set P for 50% duty cycle on an oscilloscope. Using a frequency counter, select $C_{1,2}$ for 1MHz and 1kHz. Output should be 5V into 1MΩ and 2.5V into 50Ω.

Use a ground plane and bypass the supplies close to the output stage; keep leads very short.

Thomas Korte

Hanover

Germany

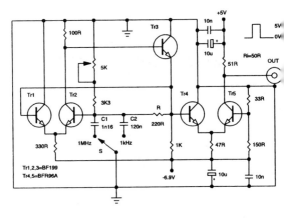

Very fast square-wave generator, with accurate levels and duty cycle, produces 2ns transition times. Use IC regulators for both supplies.

Inverting audio amplifier

I have used this amplifier in many different audio applications and found it consistent, economical and offering wide bandwidth and high gain.

DC stability with temperature is good, the necessary voltage references being derived from the two diodes and their resistors. All DC settings are easily carried out and are almost independent: v_{out} is set by Tr_1 bias resistors; cascode current by R_4; and output stage current by R_5. Values in the diagram give $V_{REF2} = -3V$, $V_{DCout} = -15.5V$, $I_1 = 0.4mA$ and $I_4 = 8mA$.

Open-loop gain is 4000 at 1kHz, falling to 2500 at 16kHz and 55 at 1MHz. Capacitor C_2 maintains

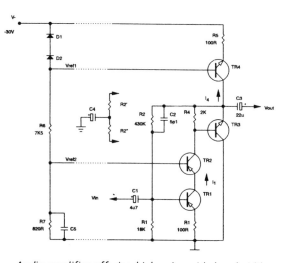

Audio amplifier offering high gain, wide bandwidth and economy, meant for use in mixers, tone controls, filters, equalisers and other 20Hz-20kHz applications. It is useful as the gain stage in a fet-input differential arrangement.

stability with feedback down to gains of less than 1.
The low-impedance output stage provides 8V RMS
at a slewing rate of 20V/µs.

For the input transistor, a low-noise p-n-p type will
give an equivalent unweighted input noise amplitude
of 0.35µV from 20Hz to 20kHz, with a source
impedance of 100Ω.

Open-loop distortion on a 5V RMS output is less
than 0.8% over the audio band, feedback to give a
gain of 5 producing a typical THD of 0.0028% at
1kHz.

As it stands, the amplifier's input impedance is low
(from 15kΩ to 50kΩ, depending on frequency) and
the two electrolytics are unfortunately needed. If the
amplifier is used as the second stage of a fet-input
differential amplifier, these problems are reduced

Vladimir Katkov
Priluki
Ukraine

Gated oscillator ignores input noise

Using a cmos 74AHCT132 quad two-input schmitt
trigger nand gate array, this oscillator will not
respond to enable signals of less than a certain width;
neither will it emit a partial pulse when the enable
signal is removed during an output low. Furthermore,
the duration of all low levels is identical – even the
first one after the enable starts.

One of the nands, U_{1B}, with C_1 and R_1, forms a
gated oscillator, its frequency depending on $R_1 C_1$.
Another nand, U_{1A}, controls the oscillator and is an

external latched gate element using feedback from the oscillator. Diode D_1 holds C_1 low when an enable high is present.

A low enable starts the oscillator. Initially, C_1 is high to hold the enable signal but, if enable goes high again before T_{min} has elapsed, the capacitor goes low, all its charge being removed. Narrow "enable" pulses are therefore ignored.

M Railesha

World Friends Design Group
Tamilnadu, India

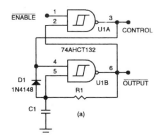

Novel gated oscillator is invulnerable to short enable signals, such as noise spikes and emits constant-duration lows.

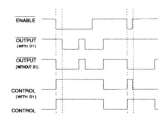

Gated oscillator with input noise rejection: input enable pulses shorter than T_{min} will not start the oscillator.

Metal detector

If the VCO output in a phase-locked loop is phase shifted and taken back to the input, the loop locks to itself and runs at whichever frequency causes 90° phase shift in the network. This principle is used to make a metal detector here, but has many applications in measurement systems.

When the search coil is within 75cm of a metal object, the VCO increases its frequency for a non-ferrous metal and decreases it for ferrous objects. Loop output on pin 7 of the *565* is compared with the pin 6 reference voltage, the long-tailed pair $Tr_{1,2}$ amplifying the difference.

My search coil is 50 turns on 50mm diameter to give an inductance of 0.5mH. The VCO frequency is around 1kHz.

Kamil Kraus

Rokycany
Czechoslovakia

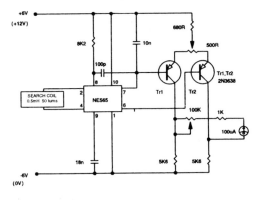

Simple metal detector using a PLL locked to itself. Search distance is about 75cm.

Fast, small-signal rectifier

For the rectification of signals of less than about 100mV at frequencies above 100kHz, the common op-amp and two diode circuit suffers from distortion and low bandwidth. This circuit overcomes the problem.

Here, the comparator detects signal polarity and connects Vin(t) in either normal or inverted form, via the Maxim *IH5141* switch, to the output. With a *TL072* as the op-amp, and an *LM360* as the comparator, bandwidth is zero to 500kHz with signals of 50mV-5V. A high-speed inverter would increase the operating frequency to 1MHz, since rise and fall times are less than 100ns.

B Jacobs, G Seehausen

Warner MM, Europe
Alsdorf Germany

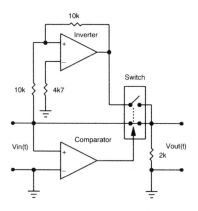

Fast rectifier for small signals avoids distortion and lack of bandwidth in usual op-amp with two feedback diodes.

Flat-top band-pass filter

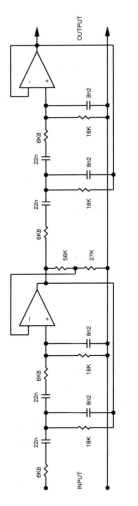

OUTPUT

8n2

6K8

18K

22n

8n2

22n

18K

6K8

56K 27K

8n2

6K8

18K

22n

8n2

22n

18K

6K8

INPUT

"Flat Top" Bandpass Filter, nominally 300 - 3.4KHz

*Exotic-looking band-pass filter for speech
frequencies is a pair of combined equal-value Sallen
and Key high and low-pass circuits, each section
using the same op-amp.*

A requirement for a band-pass filter with a sensibly linear pass-band occasioned this filter which, although having a complicated appearance, is in fact a pair of combined, equal-value Sallen and Key high and low-pass types, each section using the same active element.

Essily accessible formulae allow calculation of values, but there is one important additional requirement: to preserve the symmetry of the pass-band, time constants of the series and parallel elements must be as close as is practicable.

No awkward values are needed. The Q of each stage should be 1.306 and 0.541 respectively for an overall Q of 0.707. But, to reduce the number of components and to maintain less than 0.5dB of ripple in the pass-band, the second-stage Q is made 0.5 and the first stage Q is raised empirically to compensate. Attenuation is 24dB/octave outside the band.

Any standard op-amp, such as *5534* or *072*, is suitable.

Reg Williamson
Kidsgrove Staffordshire

Analogue A-to-D converter

This essentially analogue circuit will convert a 0-4V analogue input to a 4-bit digital output. Several such stages are cascadable and each uses a quad comparator, a quad op-amp and a *4066* quad bilateral switch.

If V_{in} exceeds $V_{ref(1)}$, the comparator output is 1 and forms the most-significant bit. V_x, which is the difference amplifier output $V_{in}-V_{ref(1)}$, is fed to the next stage to be compared in the same way with $V_{ref(2)}$. If V_{in} is less than $V_{ref(1)}$, the comparator output is 0 and V_{in} is switched straight to the next stage by the multiplexer. Only two stages are shown; the other two are identical.

The circuit finds application in a-law[1] and μ-law[2] companding.

Haydar Bilhan

EMT Electronics
Ankara Turkey

References
μ-law
1. Smith, B., *Instantaneous companding of quantised signals*, BSTJ 36, May 1957.
\A-law
2. Cattermole, K.W., *Principles of pulse-code modulation*, Iliffe, 1969.

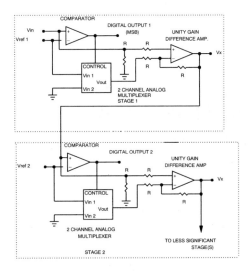

Four-bit D-to-A converter using common analogue components. Stages are cascadable.

Pulse generator

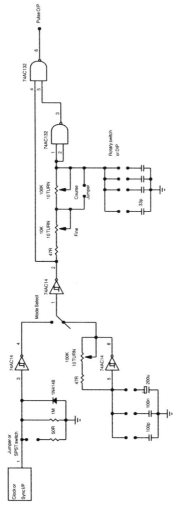

Comprehensive pulse generator providing variable-width pulses down to 5ns at up to 50MHz, or 1A 10ns pulses at 1MHz.

The pulse generator shown was designed for use in the testing of cable drivers and gives a pulse output of 5ns at up to 50MHz, or 10ns and 1MHz at 1A. Its built-in power supply will accept 5V-30V.

Clocking is by an external clock or sync. input, or by the internal clock oscillator. The two Nands and variable delay after the clock selector form the pulse generator, further shaping and inversion being carried out in the high-speed AD969 comparator, which feeds TTL outputs directly or drivers for the buffered 50Ω outputs. Pulses are also taken to the TSC430 mosfet power stage, which again provides inverted or non-inverted outputs at up to 1A.

Richard Payne
Wimbledon
London SW19

Pulse generator with independent F and M:S setting

Using a *CD4046* phase-locked loop, this pulse generator accepts independent settings of frequency and duty cycle.

Initially, the VCO in the loop holds pin six, one side of C_1 in Figure 1, to ground, C_1 being charged by a constant current whose value depends on R_3 and the frequency-setting voltage at pin nine. When the resulting ramp at point B reaches the threshold of an internal inverter, an internal flip-flop changes state, whereupon point B is grounded and point A starts to

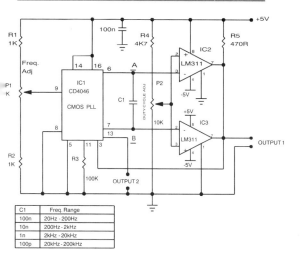

C1	Freq. Range
100n	20Hz - 200Hz
10n	200Hz - 2kHz
1n	2kHz - 20kHz
100p	20kHz - 200kHz

Figure 1. This pulse generator offers independent control of frequency and duty cycle over the range 20Hz-200kHz and 0-100%.

ramp upwards. When this too reaches the threshold of an identical inverter, the flip-flop again changes state and the cycle repeats.

Since the internal inverters are identical, the ramps at points A and B are also identical. Figure 2 shows the circuit action.

Comparison of the two ramps with a variable reference voltage in the two *LM311s* produces output 1; mark:space variation is by means of adjustment of V_{ref} and frequency setting by way of the control voltage on pin nine of the PLL. Output 2 is obtained by inverting output 1 in the PLL's phase comparator. Frequency is variable from 20Hz to 200kHz and duty cycle from 15% to 100% for output 1 and 0-85% in output 2 with a range of C_1 values from 0.1μF to 100pF.

M S Nagaraj

ISRO Satellite Centre Bangalore India

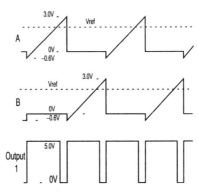

Figure 2. Waveforms in the pulse generator. Note that the ramp starts from -0.6V, not zero volts, giving a duty cycle range of 15%-100% in output 1 and 0-85% in output 2.

Product detector for AM

A CA3189 FM IF chip makes a good AM synchronous demodulator. The circuit shown provides AGC, an S-meter output and synchronous detection and only needs the RF input. Alternate half cycles of the input to pin nine are inverted by the switching waveform generated in the chip, so that unidirectional half cycles of the modulated carrier appear at pin six and simply need a filter to remove the high frequencies to leave the demodulated AM. Potentiometer RV_1 sets AGC to suit the RF stages used and RV_2 adjusts the input to avoid overload. To start with, adjust the DC at pin six to 3V on the strongest signal.

R Gough

Shenstone Staffordshire

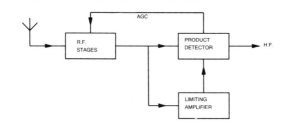

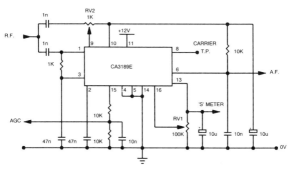

*Using a CA3189 FM IF for synchronous detection
of an AM signal.*

Switched-mode security

This circuit arrangement rapidly protects switched-
mode power supplies against overload, while
eliminating the usual protection loop and current-
sensing resistor.

It hinges on the fact that the width of the PWM
output pulse is voltage-dependent, as shown in
Figure 1. Pulses from the pulse-width modulator
arrive at the D input of a flip-flop and also trigger a
monostable, which puts out pulses that are somewhat

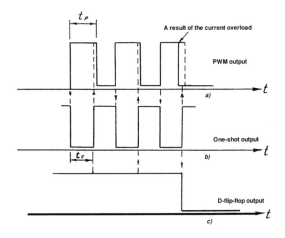

Figure 1. As the PWM's output pulse narrows with overload (a), the monostable (b) triggers the flip-flop (c), disabling the PWM. The network on S restarts the PWM at switch-on.

shorter than those from the PWM. The result is that the flip-flop's Q output is continuously high, maintaining the PWM in operation.

If an overload occurs, t_p is reduced to become narrower than the monostable pulses and the flip-flop is clocked, disabling the PWM. To restart, the circuit must be switched off and on again, whereupon the flip-flop is set by the level at S, only to retrigger if the overload is still present.

Figure 2 shows the practical circuit, using a Motorola *SG1526* pulse-width modulator with R_3 to set the power supply output voltage. Potentiometer R_4 sets the width of the monostable's pulse.

G Mirsky and A Khokhlov

Academtekh R&D Centre
Moscow Russia

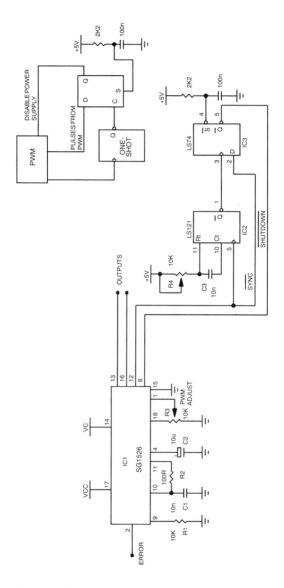

Figure 2. Practical embodiment of the Figure 1 circuit.

Resistance multiplier

To get long RC time constants without using large capacitors, use this circuit to multiply resistance values by a factor of up to 10^4. For example, a 10kΩ resistor can be made to behave like a 100MΩ component. Even the few ICs needed for the job take up less board space than a large, low-leakage capacitor.

The principle is indicated by Figures 1 and 2. Figure 1 shows a CR coupling circuit with a time constant CR when the switch is closed, voltage across the resistor falling exponentially when a negative-going step is applied at the input. If, however, the switch is open, no discharge occurs during that interval, only resuming when the switch closes; average discharge current is determined by the ratio of P_w, the time during which the switch is closed, to T, so the CR is effectively increased. The smaller pulse width in relation to the cycle time, the larger R appears.

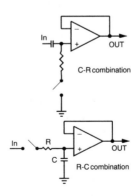

Figure 1. Switching discharge current on and off in the CR circuit effectively multiplies value of R.

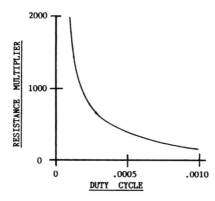

Figure 2. The normal (lower) exponential discharge curve is replaced by a stepped curve, discharging current only flowing when R is switched to ground.

Distortion caused by the steps in the discharge curve are minimised in practice by making the PRF high enough, in this case at least twice the lower cut-off frequency, which is $1/2\pi RC$. If the positions of R and C are reversed, then the PRF must be at least twice the upper cut-off frequency.

Figure 3 is the circuit diagram. A *555* timer produces switch pulses, R_1 lying in the range 10kΩ-22MΩ for a period of 350μs-4ms and R_2 between 100Ω and 100kΩ for a P_w of 4μs-4ms. Amplified output from the *555* forms the switch control, which is over-voltage-protected by the diodes; the resistor chain provides ±5V from ±15V for the *CD4066B* cmos switch IC. The *3140* is a bi-mos op-amp connected as a follower to provide a very high impedance to avoid discharging the C. Potentiometer P_1 zeroes the output. A CR circuit would be connected as in the smaller diagram.

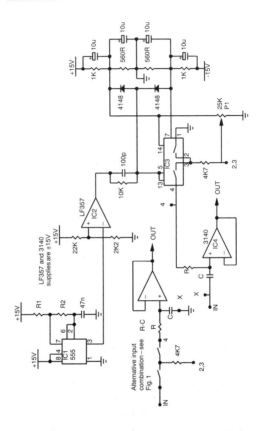

Figure 3. Full circuit diagram of the R multiplier in a CR configuration; the RC equivalent is shown opposite.

Figure 4 shows the result: with an on pulse width if 4μs, varying the PRF gives an R multiplication of between about 200 and 2000.

D A Kohl

Osseo Minnesota
USA

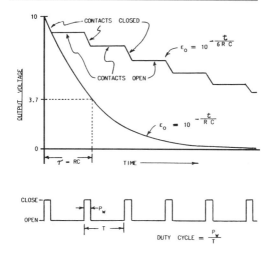

Figure 4. Varying the duty cycle with a pulse width of 4µs gives an R multiplication factor of 200–2000.

No grounds for interference

If you connect a small signal to a single-ended-input instrument and the ground potentials differ, this circuit cuts out the resulting mains interference and avoids ground loops.

A separate wire in the output cable of the signal source senses the ground potential at the instrument end, being connected to the cable shield there (inside the BNC connector with no mods needed). An op-amp adds the difference between the two grounds to the signal.

This technique can also be a useful way of preventing ground loops in audio equipment, since the 5μs filter confers a bandwidth of 34kHz; input impedance is over 100kΩ at signal frequencies. Using 1% resistors, signal-to-noise ratio improvement is, in theory, 60dB; in practice, interference simply vanished.

P C Meunier

Department of Biochemistry
University of Cambridge

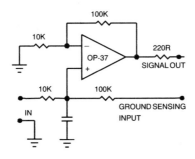

Simple circuit to virtually eliminate mains-frequency interference on small signals between a source and an instrument. Difference in ground potentials is sensed and added to signal by the op-amp. Improvement is 60dB.

Accented metronome

For a metronome with an accented beat on counts 3,4,5,6,7 or 8, three chips suffice.

Half the 555 timer, $IC_{1(a)}$, produces pulses with frequency set by the 10kΩ variable resistor and which drive the second half of the timer to give a

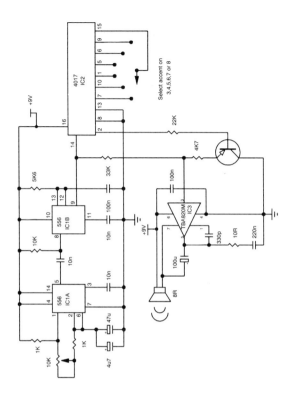

variable-frequency, constant-width pulse train at pin 9 for the Johnson counter IC_2. The pulse train also goes to the unity-gain op-amp, attenuated when the transistor is on and directly when it is off, which happens once every count cycle as determined by which Q output is switched to reset the counter. The result is an attenuated beat, followed by an accentuated count and a reset to start the count again.

D M Bridgen

Racal Radio Ltd
Reading
Berkshire.

PLL motor-speed controller

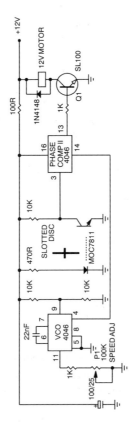

Phase-locked loop smoothly controls DC motor speed down to about 300rpm.

This is a simple closed-loop controller for fractional-horsepower DC motors, although it is suggested that larger transistors will supply larger motors.

In essence, a slotted disc on the motor shaft, moving in the gap of an optical sensor, produces pulses at a frequency f_m dependent on the motor speed and the number of slots in the disc. Pulse rate and the reference frequency generated by the PLL's VCO are compared in the phase comparator and, when f_m is less than f_{ref} the phase-comparator output goes high and Q_1 conducts, driving the motor, the reverse being the case when f_m exceeeds the reference frequency. The outcome is that motor speed is that which causes f_m to equal f_{ref} which is variable by means of the potentiometer and by altering the value of C_1.

The circuit will control motor speed smoothly at speeds down to 300rpm, assuming the disc has at least ten slots in it.

M.S. Nagaraj,
ISRO Satellite Centre,
Bangalore, India

Audible compass

Instead of graduations in degrees, this compass emits a tone which varies in frequency depending on where it is pointing – high pitch for north, low pitch for south, the difference being about an octave. It is much more stable than some other published designs.

The sensor is an RS Components *Lohet II* Hall-effect chip, whose sensitivity is increased by the addition of 60mm x 9mm ferrite rods glued to its faces. Its output is amplified by IC_1 and taken to a voltage-to-current converter having an antilog characteristic, so that equal changes in input voltage give rise to equal changes in pitch, rather than in frequency. Transistors Tr_1, Tr_2 and IC_2 form the converter, Tr_1 being the current sink for the 555 current-to-frequency converter circuitry, whose output saturates IC_3 to drive the 40mm speaker.

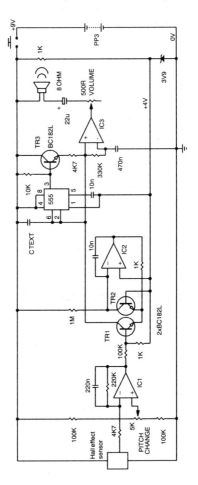

Compass gives a tone output, its pitch indicating direction. Sensor is obtainable from RS Components.

To set up, adjust the pitch control and/or board orientation to obtain 4V at the output of IC_1. Then select C to give an output frequency of about 1kHz, which is two octaves above middle C.

The three op-amps are contained in one *LM324*.

W Gough

Department of Physics and Astronomy
University of Wales
Cardiff

Phase-linear crossover

This circuit, in which a low-pass, switched-capacitor filter is cascaded with an all-pass type, gives a tunable filter with a flat characteristic – a development originally due to Lipshitz and Vanderkooy. Figure 1 shows the basics: a low-pass section and a high-pass characteristic obtained by a time delay and subtraction circuit. In practical designs, cascading a fourth-order Butterworth low-pass with a second-order all-pass equalises the attenuation and produces a phase-shift-free network, since an all-pass gives twice the phase shift of a low-pass of the same order.

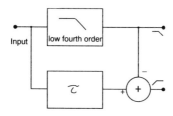

Figure 1. A phase-linear network with a low-pass section and a high-pass using time delay and subtraction.

This design uses the *ML2111* filter block; a universal, high-frequency (150kHz) dual filter consisting of two independent bi-quad switched-capacitor filters, used here in the configuration of Figure 2. Two integrators are used, in which the resistance is simulated by switching a small capacitance to give

$$V_{out} = \frac{\omega_{CLK} C_u}{sC} V_{in} = \frac{\omega_{CLK}}{s} V_{in}$$

where C_u is the switched capacitor

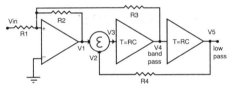

Figure 2. Configuration of ML2111 filter block, in which a resistor is simulated by a switched capacitor.

For the circuit of Figure 2,

$$V_1 = -\frac{R_2}{R_1} V_{in} - \frac{R_2}{R_3} V_4$$

$$V_5 = \frac{\omega_{CLK}}{s} V_4 = \frac{\omega^2_{CLK}}{s^2} V_3$$

$$V_3 = V_1 - V_2 = V_1 - V_5$$

Transfer functions of the universal filter are:

high-pass

$$\frac{V_3}{V_{in}} = -A \frac{s_o^2}{K_o}$$

low-pass

$$\frac{V_5}{V_{in}} = -A \frac{1}{K_o}$$

band-pass

$$\frac{V_4}{V_{in}} = -A\frac{s_0}{K_0}$$

where $s_0 = s/\omega_{CLK} = j\omega/\omega_{CLK}$

$\quad A = R_2/R_1$

and

$$K_0 = 1 + \frac{R_2}{R_3}s_0 + s_0^2$$

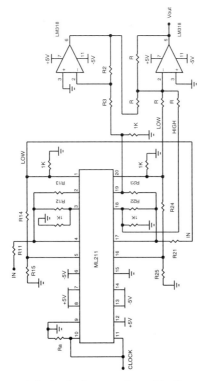

Figure 3. Full circuit diagram of tunable filter, using both halves of ML2111.

Advantages are clear: cut-off is easily set by varying clock frequency; Q can be chosen by selecting the value of R_3; gain depends on the values of R_2 and R_1 and can be set by varying R_1.

Figure 3 is the practical circuit, in which half the *ML2111* is a low-pass filter feeding the other half, the equaliser.

Kamil Kraus

Rokycany
Czechoslovakia

Three-colour bar graph

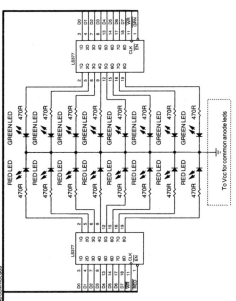

Depending on which latch is enabled, either the red set of leds illuminates or the green set. If both sets are on, an amber colour is apparent.

Using one package of green and red leds, you can get three colours — green, red and amber, the latter being a mixture of red and green used to display two sets of overlapping data.

Address lines A_0 and A_1 determine which of the sets of data from the bus is stored in the latches and whether red or green leds are on by activating port write pulses /PWR. If both A_0 and A_1 are low, both latches store the same data, all leds will be on and the colour will be amber.

Small mods to the address decoding circuit allows expansion of this idea to any number of led groups to make a bar-graph display.

J Vandana

World Friends Design Group
Tamilnadu India

Notch and high-pass filter

This somewhat unusual filter was intended to extract the 11kHz pilot tone from the Russian Gorizont (Horizon) satellite television and sound channels with a view to using it to control an *NE571* compander in its expanding mode.

At the top is a balanced bridge, with one arm taken to signal ground of a series-tuned circuit using the impedance converter, which is tunable by the 500kΩ pot to allow for circuit tolerances; for the same reason, the bridge balance is also adjustable by the 1kΩ pot. This arrangement gives a notch with a very high Q – about 40dB rejection.

The earthy end of all this goes to the virtual earth of the bottom op-amp; a high-pass filter with its maximum output at 11kHz and a steep slope. This is also adjustable because of uncertainties about the level of the 11kHz pilot tone.

Reg Williamson

Kidsgrove
Staffordshire

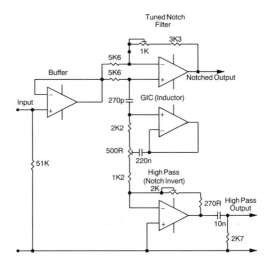

Combined notch and high-pass filter, designed to extract the 11kHz pilot signal from the Russian Horizon satellite transmissions.

VHF frequency divider

Working on an analogue principle, this divider covers the 140-150MHz range, dividing by 16 and presenting a 50Ω input impedance. It offers two main advantages over the usual logic solution: low power consumption and an input sensitivity of

around 20mV. Although this circuit operates at a
fairly low frequency, an adapted stripline version
fitted with high f_t devices could operate as a low
power divider to several gigahertz.

Division by 16 is obtained here by four divide-by
two stages, Figure 1 being the block diagram and the
practical circuit. In essence, the circuit is that of an

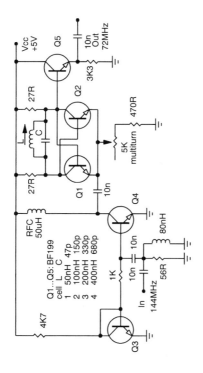

*Figure 1. Block diagram of analogue frequency
divider – basically, an oscillator set just short of
oscillation, its practical form being the emitter-
coupled "oscillator" shown. Values of C and R for
the four cells are: 50nH/47p; 100nH/150p;
200nH/330p; 400nH/680p.*

oscillator which does not start. To operate as a divider, feedback from the collector produces a voltage of the same frequency and phase as the input, the 5k trimmer in Figure 2 bringing the circuit near to oscillation and *LC* being chosen so that their resonant frequency is at the centre of the sync. range. Further cells use the input circuit shown in Figure 2.

Taking the divide-by-sixteen circuit as a whole, the minimum sync. range was 138-152MHz, output fundamental signal was about 50mV and, for inputs from 25mV to 75mV, current consumption was 30mA.

Mihai Sanduleanu

Suceava District
Romania

Figure 2. Input circuit of succeeding division cells, in which the potentiometer provides control of Tr4 bias current.

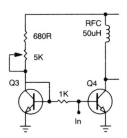

Active filter

Compared to other filter designs of this broad type, this one offers the advantages of higher input impedance, adjustable gain and adjustable Q.

High-pass, low-pass and band-pass configurations are available simultaneously, the outputs coming from V_1, V_2 and V_3 respectively.

Since $Q = R_n/R_m$ and $\omega_0 = 1/RC$, Q is variable simply by changing R_n and $_0$ by either R or C. Similarly, as the gain A of the filter is $A = (3 + R_2/R_1)$, it is also adjustable by R_1 only.

Kamil Kraus

Rokycany
Czechoslovakia

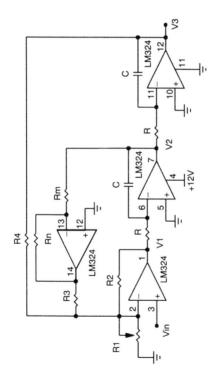

Three op-amps make up this state-variable filter,
which gives all three configurations simultaneously;
the fourth one simply allows for Q adjustment.

Zero-voltage switch

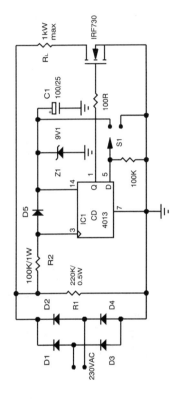

Simplified on-off switching for resistive loads taking high inrush currents.

Switching resistive loads such as lamps and heaters at zero voltage lengthens their life, but usually requires transformers and relays. This circuit uses a flip-flop and some diodes, with a mosfet.

Diode bridge D_{1-4} rectifies the mains supply, the resulting 9.1V across C_1 energising the *CD4013B* D-type flip-flop, its clock input being clipped by D_5 to V_{DD} plus a diode drop. Resistor R_1 pulls it to ground when line voltage is zero. Connecting the D input to ground or V_{DD} via the switch sets the flip-flop and the information is latched by the clock. The Q output now controls the mosfet, switching it on or off, depending on the position of the switch.

Edge triggering immunises the circuit against bounce and the zero-point switching protects the switch contacts.

M S Nagaraj

ISRO Satellite Centre
Bangalore, India

Remote keyboard for your PC

With one IC, you can add another, remote keyboard to a computer or control remote peripherals.

The *4053B* is a triple cmos single-pole, double-throw switch with very low on impedance and off leakage current. In the connection shown, the keyboard giving CLK2 and DATA2 is the default, a 0 from the remote unit's clock on CLK1 making the switch and connecting the remote keyboard for a few milliseconds. Zero and 5V keyboard supplies are common.

Ron Weinstein

Centralab Tel Aviv
Israel

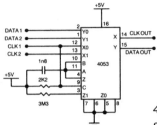

4053 cmos triple switch allows connection of a remote keyboard.

REMOTE PC-AT KEYBOARD INTERFACE

MALE WIRING SIDE 1

DATA

PC-AT KEYB CONNECTOR

Transconductance squarer

An op-amp and a dual fet combine to give an output kv_i^2 when $v_i > 0$.

National Semiconductor's *2N5452* n-channel dual fet has good matching between the two devices and low output conductance. Normally, the voltage between the fet gates and the non-inverting op-amp input is constant at V_p, the pinch-off voltage of the two transistors. V_{GS2} is $(v_i^+ + V_p)$ and i_{out} is proportional to v_i^2.

The coefficient k is adjustable by means of the $5k\Omega$ input variable, the two diodes and the $4.7k\Omega$ resistor ensuring that i_{out} does not exceed I_{DSS} and affording negative feedback should v_i^+ become greater than $|V_p|$. Output voltage must lie within the 5-15V range.

Including an absolute-value detector at the input produces a true squarer, in that either polarity of input gives the same output.

With a *741*, the circuit works at several kilohertz.

Alexandru Ciubotaru

*University of Texas at Arlington
Texas USA*

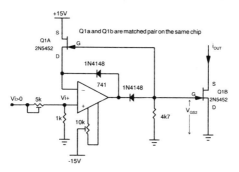

Current at the output of this simple circuit is proportional to the square of the input voltage, for inputs greater than zero. Absolute-value circuit makes it a true squarer.

Preset on time for battery equipment

This circuit was designed to switch an alarm on for a short time after a switch is momentarily made, while normally drawing no current.

Operating switch Sw_1 applies voltage to the load and to the *555* timer IC_1. Current drawn by the *555* illuminates the led, D_2, to indicate that the circuit is on and also to reduce current drain by lowering the supply voltage to the timer. At switch-on, pin 3 of the timer is high, so that $Tr_{2,1}$ are both conducting. After the momentary switch contact, Tr_2 still supplies load current until C_1 charges through R_1 – about 2s with the values shown. Capacitor C_1 eventually triggers the timer, pin 3 goes low, both transistors turn off and the circuit becomes quiescent. Diode D_1 discharges the capacitor when the load voltage collapses.

Resistor R_2 is an additional load to avoid an intermediate state in which feedback puts the timer into a linear configuration.

Steve Winder

Ipswich
Suffolk

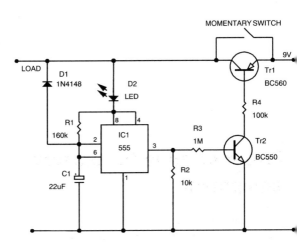

Load is energised for a short time after a switch is momentarily made. Circuit normally takes virtually no current, so is useful for battery-powered equipment.

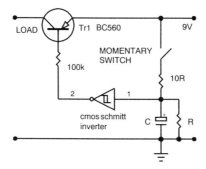

Alternative arrangement recommended by our editorial consultant.

State machine for 2.5s division ratio

In Circuit Ideas for December 1991, p.1051, Yongping Xia proposed a method of pulse frequency division by 2.5. My method uses a programmable logic device and state machine technique, thereby showing that PLDs can be used for asynchronous logic.

Figure 1 shows input and output, which are to be repeated until the starting input/output relationship is repeated, the divider then cycling in a loop.

Input conditions for every state transition are shown in Figure 2, the state chart compiled by inspection of the waveforms in Figure 1. Each of the ten states is given an unique state code and, since the operation is asynchronous, it must be a Gray code sequence in which the progression is by a change of only one code bit each time.

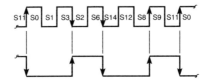

Figure 1. Input and output waveforms of 2.5 divider. Cycle repeats when starting conditions are repeated.

present(state)	if(input condition)	next(state)
	if(input condition)	next(state)
	out(output).	

Figure 2. State chart of divider system. Ten states are needed, numbered in Gray code, since the system is asynchronous.

To program the PLD, one must prepare a file to define pin functions, state code bits and logic conditions for active inputs and outputs. In the several design software packages such as Cupl, Abel and some shareware software from the manufacturers mentioned in this journal for July 1989 p.667, there is provision for defining transitions using the format:

Again, in an asynchronous system, the PLD clock input must be programmed and connected in the inactive state. A rough idea of a practical circuit is shown in Figure 3, but a manufacturer's data is needed for a working design.

J Austin

Wallasey
Merseyside

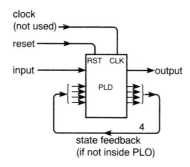

Figure 3. Basic implementation of system.

4-digit display for binary data

To present 14-bit binary signals on a seven-segment display, this circuit uses three ICs, one of them an eprom.

In the diagram, the Q4 and Q5 outputs of the free-running counter IC_1 drive the A14 and A15 addresses of the 64Kbyte eprom IC_3, input data being taken to A0-13. Output from IC_3 00-06 is a seven-segment drive signal for the display, taken via current-limiting resistors.

Outputs Q4 and Q5 from IC_1 drive a 3-to-8 decoder IC_2, whose outputs Y0-3 select one of the four dispays, since the input data chooses four different addresses. As an example, if the input is 1A4Chex., equivalent to 6732, the address is 1A4Chex. when IC_1 Q4 and Q5 are 0, IC_2 Y0=0 and the right-hand display operates. Since Y1,2,3=0, the other three are off. If Q4=1 and Q5=0, the address is

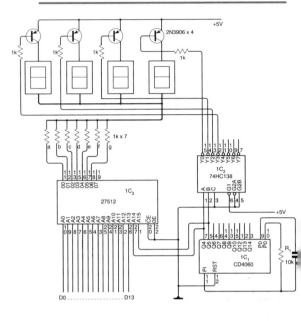

14-bit binary data connected to D0-13 appears on the seven-segment display, numbers being shown one at a time at a speed high enough to avoid flicker.

3A4Chex., Y1=0 and the next display comes on. In this way, if 1A4Chex., 3A4Chex., 5A4Chex. and 7A4Chex are programmed with the seven-segment of 2,3,7,6 respectively, the displays show these characters one by one, flicker being reduced by a high scan speed.

Maximum display is 9999 and the display is off for greater numbers.

The following QuickBasic listing generates the eprom files, files 1 to 4 storing the seven-segment forms of four digits right to left.

```
DIM N(4)
DIM SEGMENT(4)
OPEN "DATA1" FOR OUTPUT AS #1
OPEN "DATA2" FOR OUTPUT AS #2
OPEN "DATA3" FOR OUTPUT AS #3
OPEN "DATA4" FOR OUTPUT AS #4


FOR NUMBER = 0 TO 9999
    N(1) = NUMBER-INT(NUMBER/10)*10
    N(2) = INT(NUMBER/10)-INT(NUMBER/100)*10
    N(3)=INT(NUMBER/100)-INT(NUMBER/1000)*10
    N(4) = INT(NUMBER/1000)
    FOR I = 1 TO 4
            SELECT CASE N(I)
            CASE 0
            SEGMENT(I) = 40
            CASE 1
            SEGMENT(I) = 79
            CASE 2
            SEGMENT(I) = 24
            CASE 3
            SEGMENT(I) = 30
            CASE 4
            SEGMENT(I) = 19
            CASE 5
            SEGMENT(I) = 12
            CASE 6
            SEGMENT(I) = 02
            CASE 7
            SEGMENT(I) = 78
            CASE 8
            SEGMENT(I) = 00
            CASE 9
            SEGMENT(I) = 10
            END SELECT
            WRITE *I, SEGMENT(I)
```

```
   NEXT I
NEXT NUMBER
CLOSE #1
CLOSE #2
CLOSE #3
CLOSE #4
END
```

Yongping Xia

*West Virginia University Morgantown
WV, USA*

Simple, high-gain amplifier

Two extra transistors in a cascode amplifier produce a much higher gain, a greater bandwidth and a reduced output impedance.

The load R_2 resistor of Tr_4, the input transistor, has virtually no voltage across it, because of its inclusion in the amplifier made up by $Tr_{1,2,3}$; current through it is therefore practically zero. Equivalent load, and therefore gain, of Tr_4 is accordingly extremely high.

Since the $Tr_{1,2,3}$ amplifier's frequency response is wide-band, the resulting amplifier exhibits a gain of over 60dB over a bandwidth of 850kHz into 50Ω, using *BC182* and *BC179* transistors. Feedback through R_2 gives an output impedance of only a few ohms. Capacitor C_2 across the bias resistors for the input stage removes AC feedback.

I have used the amplifier in the output of an IF amplifier, in which it gave a good match to a crystal detector.

G Mirsky

Akademtekh R&D Centre Moscow Russia

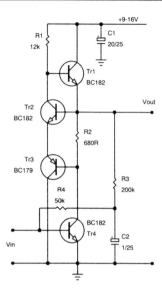

Three transistors in the load circuit of Tr4 produce high gain, wide band and low output impedance.

Voltage-to-period converter

As in traditional designs, this converter relies on a ramp technique, but in this case the flyback is initiated in a different manner and jitter significantly reduced.

The current source supplies charging current to C_1, which ramps linearly in a positive direction. As the ramp voltage reaches V_{in}, the *LM311* comparator output goes positive, the edge being differentiated by C_2R_1. The resulting pulse turns Tr_2 on, blocking the comparator at the strobe input and maintaining the output condition for a time determined by the time constant of the *CR*. It also turns on Tr_1 to discharge C_1. Ramp time T is dependent only on the input voltage and the discharge time must only be long enough for full discharge of C_1.

The relationship between T and V_{in} is adjusted by varying the value of C_1 or current source output.

Viacheslav Shkarupin

Kiev

Ukraine

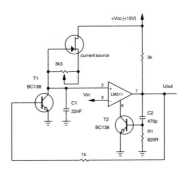

Figure 1. Positive feedback from the comparator output to initiate flyback reduces jitter in this voltage-to-period converter.

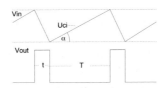

Figure 2. Slope of ramp is dependent on the value of C1 and the current source. Period T bears a linear relationship to V_{in}.

Independent on/off for long-period astable

Astable multivibrators using 555 timer ICs are flexible in that frequency and duty cycle are

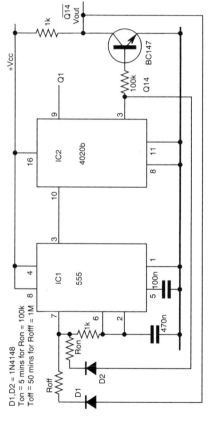

Long, independently adjustable on and off periods are obtained by switching 555 timing resistors by divider output, no large capacitors being needed.

independently adjustable but, when long and independent on/off periods are needed, large and costly capacitors must be used. This circuit avoids the problem and gives periods up to several hours.

An *4020* 14-stage binary counter divides the timer output, its Q_{14} and $\backslash Q_{14}$ outputs connecting R_{off} and R_{on} at alternate transitions to give independent on and off periods. Output Q_1 gives an indication of the output duration.

Devadoss John

Hindustan Cables Ltd
Hyderabad
India

Economical 27MHz phase modulator

This circuit phase-modulates a clock signal, a process finding application in PM and FM transmitters and in clock jitter testing. It exploits the properties of 74HC cmos gates that (a) input logic threshold increases with increasing V_{cc}, (b) propagation delay increases with increasing V_{cc} and (c) V_{cc} is allowed to vary between 2V and 6V.

Triangular modulation of 1Vpk-pk is superimposed on the nominal 4V derived from the 4.5V battery via the diode, which catches the modulating input and results in a V_{cc} varying between 4V and 5V. There is sufficient noise immunity in *74HC* and *74AC* logic to allow correct drive to the following stage with these levels.

Since the effects of varying V_{cc} on propagation delay and logic threshold are in the same sense only on falling edges of the clock input, both inverted and non-inverted clock signals are used, the timing from

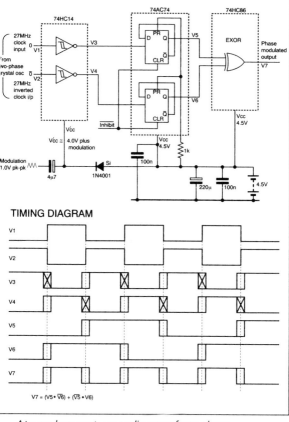

TIMING DIAGRAM

$V7 = (V5 \cdot \overline{V6}) + (\overline{V5} \cdot V6)$

At very low cost, an audio waveform phase-modulates a clock signal of up to 27MHz. Only negative-going input edges are used, since the properties of the cmos logic exploited in the circuit have a tendency to cancel on rising edges.

the rising edges being eliminated in the *74HC74*, configured as a pair of divide-by-two flip-flops. A *74HC86* Ex-Or section combines the two outputs V_5 and V_6 to provide the circuit output at up to 27MHz.

Two of the spare gates in the *74HC86* will make
the 27MHz crystal oscillator and the triangular-wave
generator was made from spare Schmitt inverters,
with a buffer, at audio frequency. A 4.5V supply is
conveniently obtained from three alkaline manganese
cells. I found it necessary to use a ground-plane
board layout.

Laurence Richardson

Hersham

Surrey

Two wire level indicator

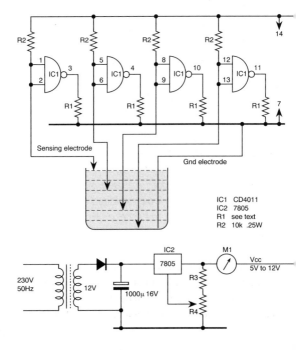

IC1 CD4011
IC2 7805
R1 see text
R2 10k .25W

Using one IC, this two-wire remote-reading
instrument indicates that water in an earthed vessel
is at or above one of four levels.

If all sensing electrodes are unwetted, all Nand
outputs are low and the meter reads virtually zero.
As an electrode touches water, its gate goes high and
the meter reads a current V_{cc}/R_1. A rising water level
causes more gates to contribute to the current.

The optional regulator and voltage-setting resistor
R_4 allow full-scale to be adjusted by shorting all the
electrodes to ground.

K N N Narayanan and C V Raman Nagar
Bangalore India.

Electronic balance for differential inputs

A true instrumentation amplifier possesses truly
differential input with good common-mode rejection
and defined input impedance. Simpler op-amp
differential inputs, however, exhibit an input
impedance that is signal-dependent, affecting CMRR
and output balance. While input impedance of the
positive input is constant, the negative input has
different input impedances for differential and
common-mode inputs. For fully equal resistor
values, impedances are:

+ve Z_{in} (balanced)   2R
+ve Z_{in} (common-mode)   2R
-ve Z_{in} (balanced)   67R
-ve Z_{in} (common-mode)   2R.

In the "Superbal", invented by Ted Fletcher and
myself in 1978, the impedance problem is solved by
applying feedback to both inputs of the first op-amp,
so that the circuit acts as an instrumentation
amplifier, but uses fewer components. Results are:

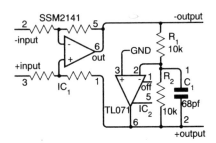

"Superbal" input stage, with feedback to both inputs to avoid input-impedance dependence on input signal.

+ve Z_{in} (balanced)   R
+ve Z_{in} (common-mode)   2R
-ve Z_{in} (balanced)   R
-ve Z_{in} (common-mode)   2R.

Using the new PMI *SSM2141*, input CMRR is in excess of 80dB with no trimming. The tolerances of R_1, R_2 and C_1 affect input impedance but not CMRR.

M Law

Acrone Ltd
Windsor
Berkshire

Phase-locked function generator

Three MSI chips will form a synchronous square/sine waveform generator. Sine output is locked to the input signal with no phase error and the square is normally in anti-phase, but the circuit is adaptable to produce zero phase error for this too.

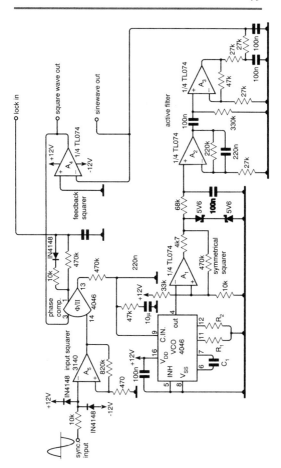

Three-chip circuit for sine and square outputs, locked in phase to an input signal.

A National Semiconductor *CD4046* VCO chip, together with op-amp A_1, generates a symmetrical square wave at a frequency determined by C_1, R_1 and R_2, values for 50Hz operation being 100nF, 330kΩ and 1MΩ. Active filter A_2/A_3 extracts the fundamental to give the sine output.

Since the filter introduces a phase delay between VCO and sine output, a further squarer, A_4, is interposed in the feedback path to the phase comparator in the *4046* to give zero phase error, "Lock in" going high when lock is achieved. Op-amp A_5, a *3140* bifet type, functions as a squarer for untidy input sync signals.

Hernán Tacca

University of Buenos Aires
Argentina

One-pin interface for an 8051

You can interface a variable resistor to a microprocessor using one pin.

An RC network is connected to an i/o-configurable microprocessor port that has either a high-value internal pull-up resistor or none at all. When the port

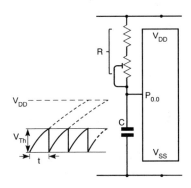

The value of R is obtained using only one pin of the microprocessor. The resistor can take the form of a thermistor, constant-current source or any resistive device.

1 program

```
RT: CLR     P0.0         :SET PORT LINE LOW
     MOV     R7,#$40      :SET DELAY TIME TO ≈180μs
AY:  DJNZ    R7,DELAY     :ALLOW CAPACITOR TIME TO DISCHARGE
     SETB    TCON.6       :START HARDWARE TIMER
     SETB    P0.0         :CONFIGURE PORT AS INPUT (HIGH-Z)
D:   JNB     P0.0,READ    :LOOP UNTIL INPUT = 1
     CLR     TCON.6       :STOP TIMER
     RTS/JMP START        :EXIT HERE OR LOOP BACK
```

is configured as an output and driven to logic 0, a short delay allows the capacitor to discharge through the port line. If the port is now configured as an input and an internal timer started, the capacitor charges through R; when the voltage on the port reaches logic 1, the timer is stopped.

Charge time is $t = CR\log_e[V_{DD}/(V_{DD}-V_{th})]$, so that timer contents are proportional to the value of R, assuming that the port takes negligible current, and if the cycle is allowed to repeat there is a sawtooth at the port pin of a frequency proportional to R.

A program suitable for the Intel *8051* is given above.

K Kirby

Burnley
Lancashire

Eight-bit data sender with no micro

Transferring parallel data from a unit that uses no microprocessor to the *RS-232C* port of a PC in serial mode requires a little trickery, interface circuits such as the *8251* not being of much help. I needed to pass data from an external system's memories to a PC.

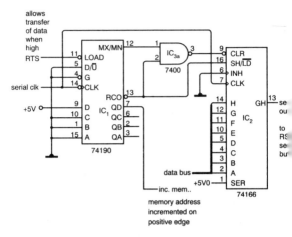

Simple circuit for transference of data to a PC from a unit with no microprocessor on board.

A *74190* up/down counter and a *74166* parallel-in/serial-out shift register, with a quarter *7400* Nand, will do the job, converting parallel data to serial 8 bits plus a stop bit, although there can be no handshake between transmitter and receiver.

I have successfully built the circuit using TTL, LS and AS devices, but have had no such luck with HC types.

Arash Ayel
Tehran Iran

Guard the vegetables without frying the cat

This electric fence generator is designed to keep out the squirrels without resorting to excessive violence.

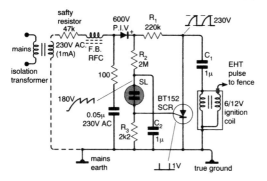

Generator for electric fences uses salvaged components and will leave the rabbits relatively intact.

Capacitor C_1 charges through the ignition coil primary and R_1 in around 250ms, C_2 charging through R_2 in a longer time. Gas discharge tube *SL* strikes at 180V, develops a voltage across R_3 and turns the SCR on. This discharges C_1 through the coil primary and the SCR goes off.

In the original circuit, only mains live and earth were used, since current is tiny. It is possible that this may cause tripping of an ELCB and an isolating transformer might be better, as well as offering safer operation – a 47kΩ resistor is used instead of a fuse to limit current to a safe value. The ferrite bead, 100Ω resistor and 50nF capacitor are for RF suppression, the capacitor possibly being taken from the same source as the discharge tube – a fluorescent-lamp starter. Capacitors C_1 and C_2 should be rated for AC or 600V DC.

J E Cronk

Prestatyn
North Wales

Resonant-loop antenna for medium waves

Resonant-loop antennas[1-3] are not quite as simple as they may appear. Induced voltage increases as the number of turns multiplied by loop area, while inductance varies as n^2, depending in a complex way on dimensions and winding method. Self capacitance and Q also must be considered.

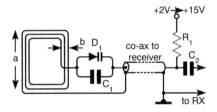

Resonant-loop antenna provides gain and interference rejection at medium-wave frequencies.

Inductance is given by Sommerfeld[4] as approximately $L = (2a\mu_o N^2/\pi)\ln(16a/b)$, where $\mu_o = 4\pi*10^{-7}$, a and b being shown in the diagram, which gives only three turns; normally there are more. Turns are spaced either in a flat coil as shown in the diagram or along the loop axis, spaced a distance $b/(N-1)$ from each other and not much more than the wire diameter.

Capacitor C_1 and the tuning diode D_1 resonate the loop and the coax. should match the receiver input impedance, which is usually 75Ω. Q will be less than $\omega L/R_0$, since the effective series resistance of the loop should be less than R_0. Make L high enough to give a sufficient Q for the selectivity needed, but not high enough to reduce C_1 to a very small value.

A rigorous test for the design is the reception in London of the 567kHz Dublin station *RTE1* in the presence of a strong local transmission on 558kHz. To use a tuning capacitance of 200pF, a loop inductance of 400μH is needed, which gives a Q of 20. With a = 1m and a 50mm spacing, the above expression gives N = 10. This was correct with 20SWG wire, the loop resonating at 567kHz with 230pF, so C_1 is 210pF and the MV1634 diode then operates somewhere near the middle of its 15-30pF range; bias voltage for the diode comes from the receiver via R_1.

Reception of Dublin was excellent with the loop in the roof space, at S9 instead of S4 with a long wire, and interference was negligible.

T H O'Dell

London W2

References
1. Marris R, *The mini-loop antenna, Radio and Electronics World, Nov. 1986, pp42-3.*
2. Millmore GW, *The long-arm loop, Short Wave Magazine, Vol. 45, No 6, Sept. 1987, p.21.*
3. Ratcliffe J, *A hexagonal loop antenna, Short Wave Magazine, Vol. 47, No 4, April 1989, pp28-9.*
4. Sommerfeld A, *Electrodynamics, Academic Press, New York, 1952, p.111.*

Feedback for lower amplifier noise

Feedback to the input of an op-amp is known to affect adversely its noise performance. In the instrumentation amplifier shown, feedback goes to the offset trimming pins rather than to the input of the difference amplifier.

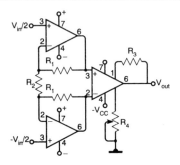

Taking feedback to trimmer points instead of input terminals preserves noise performance of the differential amp. in this instrumentation amplifier.

Voltage gain is then $(1+2R_1/R_2)k/R_4$, the value of R_4 being in kΩ. Constant k lies between 2.5kΩ and 5kΩ, depending on the type of op-amp, and should be determined by experiment.

Kamil Kraus

Roykcany Czechoslovakia

Pulsed light source

If coherence and purity of spectrum are not important in the application, this led source is capable of replacing a laser.

The amplifier formed by transistors $Tr_{1,2,3}$ accepts an input from a pulse generator and turns on the HLMP8104 led. After a short delay, determined by the rise time of current in Tr_5, Tr_6 conducts and diverts current from the led – a method of turn-off that improves noise performance.

A current of up to 10A flows in the led, at a duty cycle of less than 10^{-4}, for pulse times of 100ns to 500ns. The led was chosen for its low response time,

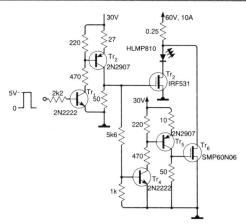

Simple led source can produce 1μJ pulses of light and will replace a laser in measurements where coherence is not needed.

even when overdriven. Peak light intensity is several W/cm^2, equivalent to about 1μJ. After up to $2*10^6$ pulses at this level, light intensity reduces to about half the starting value.

Luigi Schirone

Università "La Sapienza"
Rome Italy

Voltage-tolerant temperature probe

In the presence of reverse voltages up to 44V, this circuit is an improvement on the common type using only the Analog Devices AD590F temperature sensor, which can tolerate up to 20V reverse voltage.

Figure 1 shows the arrangement, which uses a bridge rectifier to allow the sensor to conduct for either polarity of power source and which protects the AD590F in the event of a change in polarity. Output is either current (1μA/K) or voltage (1mV/K), depending on the presence or absence of the resistor R_1.

Miss Railesha

Tamilnadu India

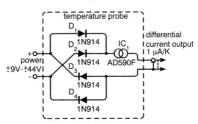

Figure 1. Addition of bridge rectifier to AD590F temperature sensor allows use of up to 44V positive or negative supplies without danger to the sensor.

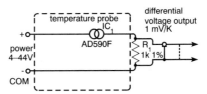

Figure 2. Standard AD590F tolerates only up to 20V reverse voltage.

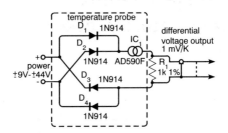

Figure 3. Bridge rectifier circuit with voltage ouptut.

Overcurrent protector

Placed between a power source and its load, this circuit protects the power source from excess current drain. It limits current surge at switch-on and acts as a circuit-breaker if a short or circuit malfunction causes current to exceed a preset limit in normal operation.

With the switch on, the series transistor Tr_1 is biased on via R_5 and the zener D_1, bias being set to give the required maximum output current of 100mA or less. The load receives virtually the entire rail voltage.

Excess current flowing through R_1 brings Tr_2, and therefore Tr_3, into conduction, diverting base current

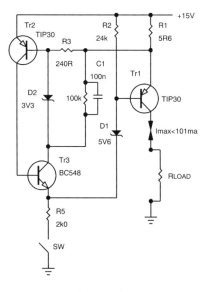

Overcurrent protection circuit prevents power-supply damage caused by switch-on surges and acts as a circuit breaker in the presence of a short-circuit.

from Tr_1 and limiting output current for a short time, as in a switch-on surge. If the high current persists, C_1 charges up to the voltage of zener D_2, which connects Tr_3 collector to Tr_2 base and produces an avalanche effect. All the available bias current is now diverted from the output transistor and no output current passes.

The time to complete cut-off depends on the time constant C_1R_5.

N I Lavrantiev

Schiolkovo
Moscow Region Russia

Function generator is digitally programmed

By means of a simple modification, the function generator put forward by R W J Barker in *Circuit Ideas* for June, 1991 becomes programmable from a digital input word.

Figure 1 shows the original circuit, which is a ring oscillator producing approximate square, triangular and sine waves at x,y and z respectively, its frequency being determined by R_1C_1. In Figure 2, the resistance seen between (a) and (b) is $R_{ab}=1/g_m$, where $g_m=I_{ABC}/2V_T$, the transconductance of the two transconductance amplifiers, I_{ABC} being the automatic bias control current and V_T the thermal voltage. Since the value of R_1 controls the frequency of oscillation, replacing R_1 with this circuit allows linear frequency control by variation of the input current.

Adding a digital-to-analogue converter, as shown in Figure 3, produces a digitally-controlled, variable

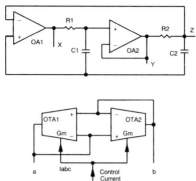

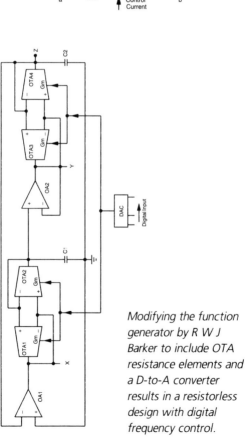

Modifying the function generator by R W J Barker to include OTA resistance elements and a D-to-A converter results in a resistorless design with digital frequency control.

frequency function generator, which has been realised using *LM13600* operational transconductance amplifiers and *741* op-amps.

Muhammad Taher Abuelma'atti and Sulaiman Al-Gharbi Al-Sayed

King Fahd University of Petroleum and Minerals
Dhahran, Saudi Arabia

Precise power output stage

When a series regulator must both source and sink current, or if the standing current in an audio power output stage must be accurately set independently of temperature, then this circuit is one solution.

Since top and bottom circuits are identical, apart from polarity, the top half will be described. When quiescent, the current mirror $Tr_{1,2}$ has a voltage between the emitters which depends on standing currents according to $V = V_T \ln I_1/I_2$, where V_T is the temperature voltage kT/q of 26mV at room temperature. If $I_1 = 10I_2$, $V = 59$mV at 25°C ambient and setting $R_{5,6}$ at 1Ω puts the standing current in Tr_5 at 59mA, independently of junction temperature, the values of $R_{5,6}$ and the ratio I_1/I_2 being adjusted to suit one's needs.

If $R_1 = R_2$, small-signal input impedance is

$$R_{in} = \frac{1}{2} \frac{g_{fc} R_1 R_L}{1 + g_{fc} R_L} = \frac{R_1}{2},$$

when $g_{fs}R_L > 1$, where g_{fs} is the transconductance of $Tr_{5,6}$, although replacing the resistors by current sources will increase that. Output impedance is $R_7/2$. Match the mirror pairs to avoid errors and to prevent possible thermal runaway.

Terence S Finnegan

Carlisle

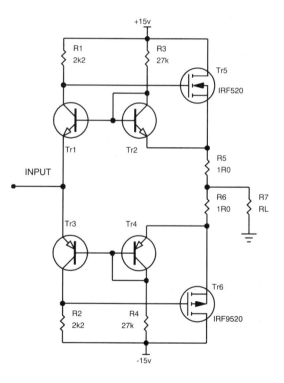

Current mirrors in this power output stage, which sources and sinks current, allow accurate setting of standing current.

Fast full-wave rectifier

Loosely based on a design by Lidgey and Toumazou (EW^3, November 1987, p.1115), in which current mirrors sensed the supply current of op-amps, this circuit uses a *MAX435* wide-band, differential-output transconductance amplifier to give full-wave rectification of signals up to 250MHz. Output is $4Z_L/V_{in}$ for the *435*. You could also try the Burr-Brown *OPA660*, which offers 700MHz-plus operation.

Peter May

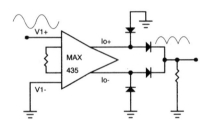

An up-date on a design by Lidgey and Toumazou, using a MAX435 or a Burr-Brown OPA660 for very high-speed rectification.

Near-field probes for EMC testing

Before spending money on having a new product assessed for its EMC, it might be advisable to check roughly on its noisiness while still in development. The diagrams show two probes for near-field "sniffing": an electrostatic probe and an electromagnetic type.

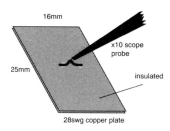

Near-field electrostatic probe allows low-cost testing of prototype equipment for electromagnetic compatibility.

The former is a thin plate of copper or tinned steel measuring about 16 by 25mm and having a hoop of 20swg wire soldered to it so that an oscilloscope probe can clip onto it. The plate is insulated with tape, since it is used near live circuits. A 25mm length of wire carrying a 4Vpk-pk, 31kHz square wave gave a 5mV pk-pk oscilloscope deflection at a distance of 10mm. Holding the plate edge-on to a PCB track gives the best signal.

As a less expensive alternative to the Tektronix *Alternating Current Probe* with the jaws open which, as the table shows, worked reasonably well, my solution is 10 turns of enamelled copper wire at 20mm diameter. On signal transitions, this gives triangular spikes about 100ns wide, which trigger most oscilloscopes; loop currents of 2mA pk-pk are

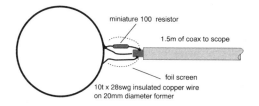

Electromagnetic probe.

source detector	20mA in loop	40mm dia.	60mm dia.	decay time
Tek. P6021 probe (open)	switch to 2mA/mV	1.5mV	1.0mV	5µs
	switch to 10mA/mV	0.3mV	0.2mV	25µs
10t on 20mm dia with 100Ω		40mV	25mV	100ns spike

visible at 5mV/div. The 100Ω resistor gives a slightly under-damped response and a larger signal than with 50Ω.

The table shows measurements made with the detector coil at the centre of the source loop. If a spectrum analyser or a fast, sensitive oscilloscope is used, the number of turns can be reduced to give a truer spectral response.

C J D Catto

Elsworth
Cambridgeshire

Independent m:s adjustment for wide-band pulse gen

This circuit delivers square waves and rectangular waves with a mark:space ratio of 10-90% at frequencies from 1.2kHz to 2.7MHz

Frequency generation comes from the voltage-controlled oscillator IC_1 and associated components, the output of which is adjustable from 12kHz to 27MHz by means of the potentiometer.

Johnson counter IC_2 provides a Set pulse from the Q0 output to the SR flip-flop IC_3, the corresponding Reset pulse coming by way of the selector switch from outputs Q1-Q9, output frequencies being $^1/_{10}$ of the input from IC_1, as is the square waveform from carry output C0.

W Dijkstra

Waalre, The Netherlands

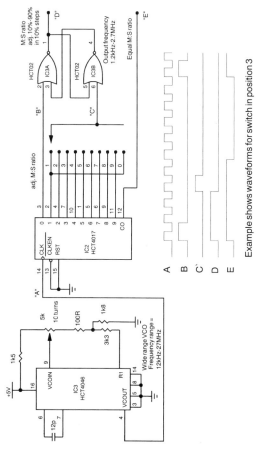

Three ICs form a 1.2kHz-2.7MHz pulse and square-wave generator having a mark-to-space ratio adjustable in 10% steps from 10% to 90%.

Stepper motor control

Two chips, a universal shift register and a darlington transistor array, form a stepper motor controller with no visible discrete devices; the free-wheeling winding diodes are in the array.

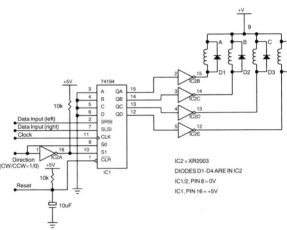

This stepper motor controller uses no discrete semiconductors, the diodes being contained in the transistor array XR 2003. The circuit controls motors taking up to 500mA per winding.

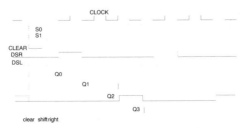

Waveforms seen in the circuit for clockwise motor rotation.

Pulses into the data shift right/left inputs of the *74194* universal shift register produce logic sequences for both directions of motor rotation, depending on the polarity of the direction control signal to the S_0 and S_1 inputs; the input data is sequentially output from Q_{0-3} or Q_{3-0} at each positive-going clock pulse.

Each device in the *XR2003* seven-transistor array handles a 500mA continuous collector current at up to 45V and devices may be paralleled. The inverter on the direction input to the *74194* is part of the *XR2003*.

V Lakhshminarayanan

Centre for Development of Telematics
Bangalore
India

Programmable instrumentation amplifier

Gain of this three-op-amp amplifier is given by $A=(1+2R/R_x)$, x being the value of one of the resistors in the *IH5070*. Selecting this resistor by the A_0-A_2 lines produces a programmed-gain amplifier.

The *TAB1042*, as well as being a conventional op-amp, is also an analogue switch, shutting down when no bias current goes to pin 8. A timer feeds pin 8, so that the operating time is programmable from about 1s to 24h.

Kamil Kraus

Rokycany
Czechoslovakia

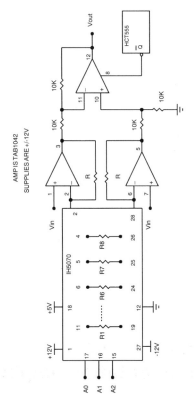

Three op-amp instrumentation amplifier, programmed in time range and gain, which is set by selecting one of the IH5070 resistors by the A0-A2 digital input.

Three-phase indicator

Phase sequence in a three-phase supply is shown by red and green leds.

Phase 1 drives a red led, is rectified to provide 5V for the gates and, after the *74HC14*, emerges as 3ms

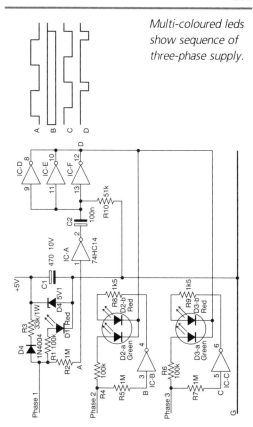

Multi-coloured leds show sequence of three-phase supply.

negative-going pulses. Phase 2 drives a green led and is inverted to drive the red led, which is also controlled by the 3ms pulses from point D. Phase 3 is the same as phase 2.

If the phases are in order, phase 2 red led is on and phase 3 red led off so that, since the pairs of leds are in the same packages, phase 2 shows yellow and phase 3 green. If phases 2 and 3 are out of order, phase 2 is green and phase 3 yellow.

Yongping Xia

Torrance California USA

Microprocessor/analogue voltmeter

Originally intended as a microprocessor interface for DC, or average, peak and RMS AC voltage monitoring under software control, this will also serve as the core of an analogue voltmeter for DC and RMS measurement. Full resolution from the microprocessor's A-to-D converter is retained for either polarity, a differential type being unnecessary. The circuit provides for automatic polarity recognition and either automatic software or manual range setting.

TL074 op-amps 1 and 2 provide inverted and non-inverted inputs to two fet switches in the *4066*, scaled by R_y/R_x. Op-amps 3 and 4 are comparators, which recognise polarity and drive the fet gates in such a way that the output capacitor charges or discharges unidirectionally. The three-coloured led across their outputs shows red or green for polarity and yellow for AC.

For an analogue indication, a 2.5V meter across the output capacitor will serve and manual range-resistor selection eliminates the *4051* multiplexer, which is software controlled when the circuit is used as a microprocessor interface. With R alone in series with the meter, DC or an average-value reading results; to make it read RMS, the 10Ω resistor across it increases the reading by 1.11.

On a practical note, the input limiting diodes must be shielded against light and high $R_{x,y}$ values require care with the layout.

H Maidment

Wilton
Salisbury Wiltshire.

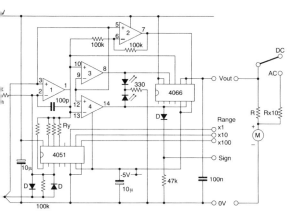

Interface circuit for voltage monitoring by microprocessor or analogue voltmeter to indicate RMS or DC of either polarity.

Measuring voltage in interference

Although integrating A-to-D converters can reduce the effects of mains-frequency interference when measuring small, low-frequency voltages, other types of interference, the periods of whose fundamental and harmonics are not multiples of A-to-D integration time, still cause trouble. This circuit alleviates this situation.

In essence, a sample of input signal is taken at the moment the interference is at zero, so that the remaining input is purely signal. The A-to-D holds the sample until the next EMI-zero sync. pulse one period of interference later.

Figure 1 is the basic idea when the only interference is at mains frequency. Signal plus noise is amplified and passed to the converter, which only produces a

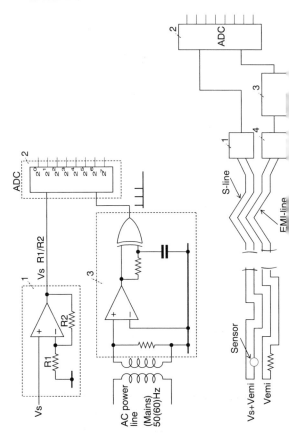

Figure 1. Interference-reducing circuit for low-frequency voltage measurement in the presence of EMI of up to 100 times signal amplitude.

conversion when the mains-frequency sync. pulse appears from the logic circuit. Figure 2 shows the effect.

In the case of random-frequency interference, it is preferable to use a dummy sensor instead of the step-down transformer, reproducing the measuring sensor's

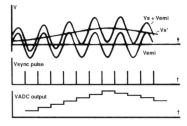

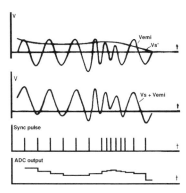

Figure 2. Sync. pulses initiate A-to-D conversion when EMI is at zero, leaving pure signal.

resistance, capacitance and inductance, connected by the same length of line and taken to a similar amplifier. The rest of the circuit is the same as that of Fig.1. If the interference is at a relatively high frequency, phase shift between the sensor and dummy sensor inputs should be negligible. A *TDC11007* converter, with a conversion time of 0.15µs, and an *AD509* amplifier worked well in the prototype.

To use a slower converter, the simple capacitive storage circuit in Figure 3 is feasible. Sync. pulses open the switch and begin the conversion, the converter's ready pulse closing the switch to allow a further reading at the next sync. The capacitor discharge time must not exceed one period.

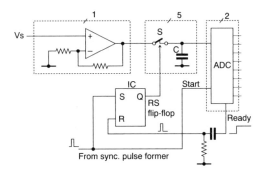

Figure 3. Simple storage circuit to allow slower A-to-D converters.

At low signal frequencies, this circuit is valid when EMI is up to 100 times greater than signal voltage.

N T Lavrentiev

Moscow
Russia

555 has high mark:space

An astable flip-flop using a 555 timer in the usual circuit will not operate with a duty cycle of less than 50%. Diode D_1 shown in the diagram overcomes the problem; it allows a duty cycle from very near zero to virtually 100% while keeping a reasonably constant frequency.

$$T_{(on)} = R_1C \left(\ln \frac{3 - k}{1 - 3k} - \ln \frac{3 - k}{2 - 3k} \right)$$

and $T_{(off)} = 0.695R_2C$, where

$R_1 = R_a$ + resistance of upper leg of pot.,
$R_2 = R_b$ + resistance of lower leg of pot.,
k = diode forward drop/supply voltage.

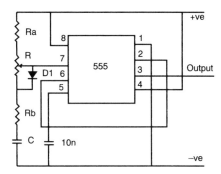

Adding the diode to a 555 multivibrator circuit allows a much larger range of duty cycles – from virtually zero to 100% – with a fairly constant frequency.

If the diode drop is very small compared with the supply voltage, k can be neglected, giving $T_{(on)} = 0.695\, R_2 C$, duty cycle $R_2/(R_2 + R_1)$ and frequency = $1.44/C(R_1 + R_2)$.

Cyril WW Palihawadana

Sana'a
Republic of Yemen

Coarse/fine D-to-A audio attenuator

Two 8-bit digital-to-analogue converters are combined in this circuit to obtain an effective resolution of 14 bits, offering coarse and fine control.

Ignoring $R_{3,4}$ in the diagram, $R_{1,2}$ will combine the output currents of the *AD7528* dual A-to-D converter. Resistors $R_{1,2}$ attenuate the input by the required amount; for example, if $R_1 = 63R_2$, the reference voltage of *DAC B* is 1/64 that of *DAC A* and the range of *DAC B* is 4 least-significant bits of *DAC A*. A V_{in} of 2.56V gives a *DAC B* reference voltage of 40mV and an LSB of 156μV.

Problems exist, however, with the basic arrangement. The output would not be monotonic, since the ratio of $R_{1,2}$ would need to be perfect and the input offset voltage of the output amplifier would contribute a code-dependent noise gain term. Additionally, there is the loading of *DAC B* input R

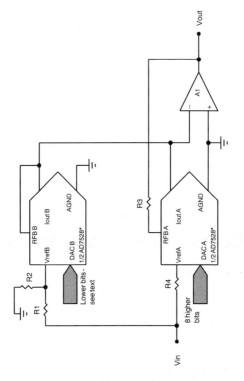

Two 8-bit DACs combine to make a 14-bit system for audio attenuation.

across R_2, although this can be eased by making the resistance a low value. Then again, the temperature coefficients of $R_{1,2}$ will not match that of the internal ladder.

All this is avoided by the inclusion of R_3 in series with the internal feedback resistor. Making this value $R_{1,2}$* means that *DAC B* reference voltage is only a function of the ratio R_1:R_2. Resistor R_3 should also have a temp.comp. similar to that of $R_{1,2}$. Resistor R_4 compensates for R_3.

Output voltage is now

$$V_{out} = -D_A V_{IN} - D_B V_{IN}[R_2/(R_1+R_2)],$$

$D_{A,B}$ being fractional representations of the input code N in decimal; that is, $D_A = N_A/256$.

*Brokaw, Paul. Input resistor stabilises MDAC's gain, *EDN*, Jan. 7, 1981, p.210.

John Wynne

Analog Devices
Limerick Ireland

Current source has 170V voltage compliance

Using a three-terminal adjustable voltage regulator chip, the *TL783C*, this circuit arrangement will drive switchable currents from 25mA to 100mA into a load varying between a few ohms and up to 1.7kΩ. *TL783C* maintains a nominal and stable 1.25V between OUT and ADJ pins, which is used to drive current through resistors switched to give 25,50,75 and 100mA output to the load. Current into the ADJ pin is a very stable 60μA and, since this is referred to the output, it sets a minimum output current of 60μA and therefore a maximum load resistance. The rest of the circuit is designed to manage this.

In the low position of S2, maximum input/output voltage and dissipation are no problem, but an increasing load resistance reduces $V_{in}–V_{out}$ and $V_{out}–V_{adj}$. Comparator IC_{1b} turns on Tr_3 and the led illuminates to show out-of-range at about a 1kΩ load resistance at 100mA.

With S2 at high, a load resistance of less than 700Ω or a short circuit would destroy the circuit, although the load can go up to 1.7kΩ. In this situation, the 3W

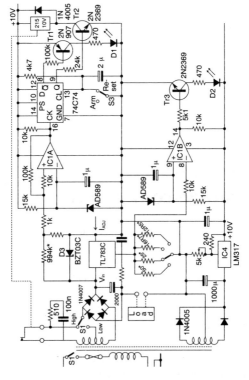

Voltage regulator chip used in a 25-100mA current source with voltage compliance up to 170V.

zener across the *TL783C* limits $V_{in}-V_{adj}$, the schmitt IC_{1a} detecting the limiting voltage and clocking the flip-flop. Both Tr_2 and Tr_3 turn on, the relay disconnects the input and led D_1 lights. A reset to the flip-flop reinstates the circuit when order is restored. Voltage compliance in the low mode is 110V; taking into account both modes, 170V.

J J Meyer
ITODYS
Paris.

Coaxial-cable tester

Three leds indicate the condition of a coaxial cable: whether it is short-circuit, open-circuit or good. A constant-current ring-of-two circuit, zener diodes and a couple of flip-flops in a 7474 comprise the circuit.

The current source works as follows. Initially, current flows through the 680 resistor to Tr_1 base; as the supply voltage increases beyond $2V_{be}$, voltage across R_2 increases and supplies base current to Tr_2. With a further increase, Tr_2 conducts more heavily, diverting current from Tr_1 and keeping a constant current between A and B. In this application, the source supplies 5mA ($I_1 = V_{be}/R_2$ and $I_2 = (V_{R1}-2V_{be})/R_1$).

To test a cable, connect it by BNC connectors or other suitable types across X and Y and reset the flip-flops by the switch. If a good cable is under test, 5mA flows through the cable to the 3.3V zener D_5 which, being a lower-voltage type than D_1, prevents it from conducting. Since the Q\ flip-flop outputs are high when reset, Tr_4 is gated on and lights the green led D_8.

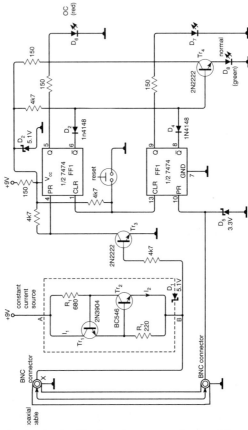

Simple circuit affords go/no-go test for coaxial cables.

An open-circuit cable results in current being diverted through D_1 and Tr_3, setting flip-flop 1 and lighting the red led D_6.

Short-circuit cables pull the preset input of flip-flop 2 to ground, setting it and lighting the amber led D_7.

V Lakshminarayanan

Centre for Development of Telematics
Bangalore India

Car intruder alarm

Three minutes after leaving the car, this alarm becomes effective and, on re-entering, provides about fifteen seconds grace in which to disable it before sounding for ten minutes.

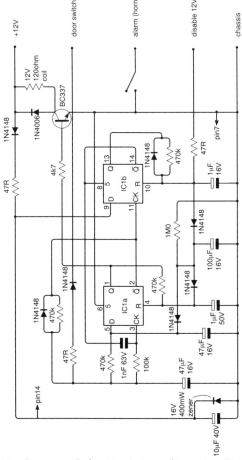

Car alarm sounds for 10 minutes, after giving 15 seconds grace to disable it.

Existing courtesy-light switches are used to activate the alarm. Each half of the dual D-type flip-flop is a monostable, regeneration being provided by C_6R_6, once the circuit has been triggered and as long as the voltage on the first flip-flop D input remains above the threshold; in the quiescent state, both clock and D input always return to a positive state. Applying a DISABLE signal forces both RESET and D inputs positive.

The components R_1, D_5, R_4 and C_1 determine the time delay before the alarm becomes effective after the car is vacated and R_2C_2 give the delay on re-entry and the doors closed, which is to say after the earth through the door switch and R_8D_3 has been removed. Alarm duration is set by R_3C_3 and the duty cycle by R_4C_4 and R_5C_5.

Once the alarm is triggered, the door switches have no effect on the alarm, which resets itself 30 minutes after it has first sounded. The disabling supply can be taken to the auxiliary pole of the ignition switch or a separate hidden switch.

H Maidment

Wilton
Wiltshire

Motion-direction detector

Depending on which of the inputs x_1 or x_2 receives a logic 1 first, the outputs produce $F_1F_2 = 10$ or $F_1F_2 = 01$, thereby indicating the direction of motion of an object breaking an infrared beam. For the sequence of inputs $x_1x_2 = 00$-10-11-01-00, F_1 will be high and for the progression 00-01-11-10-00, F_1 is low.

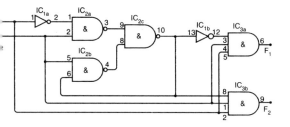

Figure 1. Logic circuit of motion detector, where state of outputs indicates direction of motion of object breaking beam.

Figure 2 shows the application, in which the two beams are used to indicate motion. When F_1 is high, beam A has been obstructed first and the motion is therefore from left to right. Such an arrangement can be used where separate counts are needed of objects moving in opposite directions, the infrared beams being replaced by other types of sensor, if necessary.

M.Kumaran

University of Keele
Staffordshire

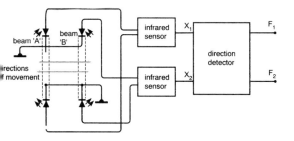

Figure 2 shows application of detector. Beams may be replaced by switches.

High-voltage controller using OTAs

Needing only $100\mu A$ differential drive, the circuit shown provides an output voltage change of 100V in 1% increments.

Drive current is supplied by an operational transconductance amplifier IC_2 and power gain by three mosfets. Transistor Tr_1 is a source follower, its gate voltage being set by the voltage drop across about $1M\Omega$ of current-sinking load, which is split between the two other mosfets to keep majority current power dissipation within limits. The combination forms a controllable power zener to set the load voltage across R_L. Transistor Tr_3's gate threshold voltage is above the minimum compliance level for the current mirror at IC_2 pin 12.

Bias voltages of +5V and +10V were needed for other sections of the system, so the input divider is calibrated by the variable resistor for nine 200mV steps across matched 100Ω resistors, contained in a thick-film package. Op-amp IC_1, with a gain of 0.1, accesses the same tapping points to produce two-decade resolution and good tracking at the OTA differential inputs.

This first OTA is a linear differential voltage amplifier which drives a $20k\Omega$ load from +5V; linearising diode bias at pin 1 causes the 2V input to appear as a current to the load resistor. The resulting voltage drop, buffered by the second op-amp, becomes a +5V to +7V swing on a constant 3V, so that the second OTA receives a 2:1 voltage excursion at its inputs.

Diode bias at pin 1 allows proportional current sinking into pin 12 to determine the 2:1 output

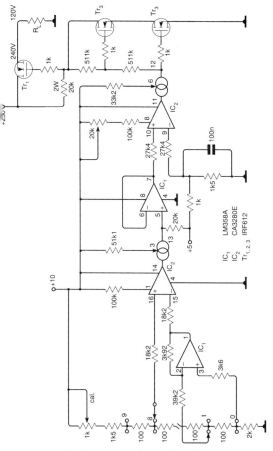

Current-mode design permits both a wide-range voltage amplifier and large-differential voltage-to-current conversion using the same package in this voltage controller.

swing. The 20kΩ variable resistor adjusts range and magnitude.

John A Haase

Fort Collins Colorado USA

4 X 4-bit parallel binary multiplier

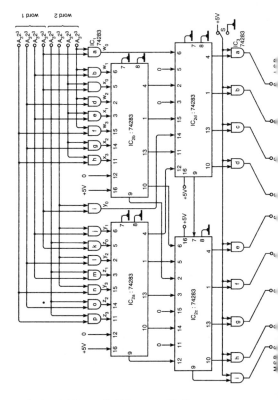

A 4 x 4 bit parallel binary multiplier.

Large computers and signal processors need to perform high-speed multiplication and do this by using arrays of gates, half and full adders. The arrangement shown is, so far as I am aware, new and is a four-bit parallel binary multiplier using a four-bit full adder, the 74283.

If two, four-bit binary numbers, A3-A0 and B3-B0, are to be multiplied, the partial products are:

W0 = A0B0; W1 = A1B0;
W2 = A2B0; W3 = A3B0
X0 = A0B1; X1 = A1B1;
X2 = A2B1; X3 = A3B1
Y0 = A0B2; Y1 = A1B2;
Y2 = A2B2; Y3 = A3B2
Z0 = A0B3; Z1 = A1B3;
Z2 = A2B3; Z3 = A3B3.

Add the partial products in W and X and those in Y and Z, then add these two partial sums to obtain the product; binary place values of the partial products must not be altered during the addition.

In the diagram, the array of And gates in IC1 provides partial products W,X,Y and Z, W + X and Y + Z being produced separately in the full adder IC2a,b. Full adders IC2c,d add these partial sums, the output of the second array of And gates in IC3 giving the final product of the two four-bit numbers.

This method is easily extendable to cope with eight-bit binary numbers.

PR Narayana Swamy

Kingswood
NSW Australia

Sound sampler filter

Using an 8-bit sound sampler board with Atari, Archimedes or Amiga computers for sound analysis in physics and biology requires a sample rate of at least twice the highest signal frequency – around 20kHz. Some samplers possess a simple low-pass filter rolling off at about 16kHz, but do not remove

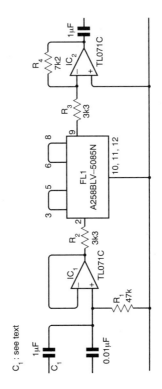

Low-pass filter is steep enough to allow sound sample rate of twice signal frequency, with attenuation of better than 50dB at 20kHz.

enough of the signal above 20kHz to allow a sample rate of 40kHz. The diagram shows a low-pass filter to give the necessary performance using an inexpensive LC element, which is intended for Nicam systems.

Buffering is provided by R_1, C_1 and IC_1, R_2 matching the output to the filter input impedance. Resistor R_3 matches the output to IC_2 and, together with R_4, sets the gain of the circuit, which is 0dB overall when circuit losses are taken into account. Components C_1 and R_1 form a high-pass filter, rolling off at 3.4Hz or 340Hz, depending on which value is used.

Prototype performance was within 1dB between 700Hz and 14.5kHz, with roll-off points at 330Hz and 15.2kHz; above 19.7kHz, signals are at least -51dB. The filter is a Toko part, an *A258BLV-5085N*, from Cirkit and Maplin.

L May

Rochdale
Lancashire

Multimeter as frequency meter

A digital multimeter makes a good frequency meter with the addition of the circuit shown.

In principle, the average direct voltage of a rectangular waveform depends on the ratio of a fixed-width part of the waveform to the variable width of one cycle.

In the diagram, the input signal, after any prescaling in $IC_{3a,b}$, and IC_{4a}, generates a very narrow pulse at the output of the second Nor, which is used to reset 14-stage binary counter IC_1. The counting then proceeds under the control of a built-in crystal-controlled oscillator until Q_{10} becomes 1, whereupon the oscillator is made to stop. The crystal ensures that the interval between reset and stop is stable and accurate. The width of the interval is $T_c = 1/14.3 \times 512 = 35.8\mu s$ and the cycle time is $T = 1/f_{in}$, so that the average output direct voltage is

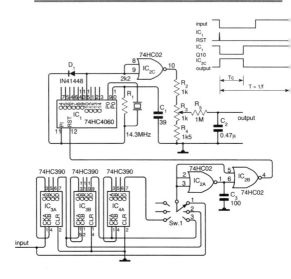

Low-pass filter is steep enough to allow sound sample rate of twice signal frequency, with attenuation of better than 50dB at 20kHz.

$$V_0 = V_{dd} \times T_c / T = V_{dd} \times T_c \times f_{in},$$

where V_{dd} is the voltage of the power rail. Since V_{dd} and T_c are constants, output voltage is proportional to input frequency, R_5 and C_2 smoothing the square wave to give the direct voltage to drive the voltmeter.

Switch positions 1 to 4 select the divider outputs to give full-scale ranges of 19.99kHz to 19.99MHz. To calibrate, adjust R_3 to give a 1.9V direct-voltage output when the input is 19kHz.

Yongping Xia

West Virginia University
Morgantown WV
USA

Motion direction detector

I suggest that a simpler approach to detecting the direction of motion of an object, one method of which was put forward by M Kumaran in July, is to use half a *74LS74* D-type flip-flop, as shown in the diagram, which performs exactly the same function.

A Clark

Boldon Colliery
Tyne and Wear

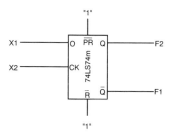

A simpler motion detector.

D-to-A converter current booster

An increase in output current from the popular *DAC 08* D-to-A converter is obtainable by means of this circuit, the current accuracy being better than 1% and offset current less than 0.1%.

Current gain is 16 precisely and output voltage lies between –8V and +25V. The only thing to watch is the likelihood of oddities at large transitions — all 0s to all 1s, for example. The relatively low slewing rate of the LM308A is responsible for this and imposes a limit on clock frequency of 15kHz if such an eventuality is at all likely.

Alexandru Ciubotaru

Galati

Romania

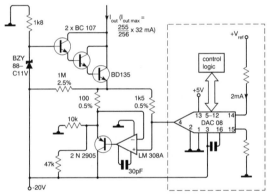

Precisely sixteen times the 2mA output current from a DAC 08, with a low current offset error.

Sine waves from a 4046 VCO

AM detection and waveform generation are two of the expanding number of uses for the phase-locked loop IC, the one in question being the 4046 cmos type. Quadrature sine or triangular waves are derived from the VCO timing capacitor in a similar manner to that described in ref.1.

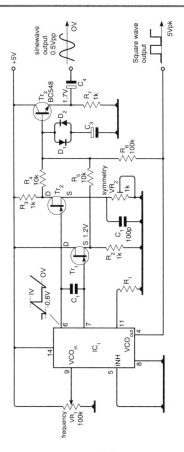

Function generator using a PLL produces square and sine or triangular waves from 20Hz to 18MHz.

Voltage across the timing capacitor C_1 is a linear differential sawtooth having a constant peak voltage with frequency. The two fets make up a high input-impedance differential pair with unity gain and over 10MHz bandwidth, which maintains linearity at low charging currents and at high frequencies up to 18MHz. Summing the two fet outputs in Tr_3 gives a triangular wave and a small square component, which is removed by taking some of the square at the VCO output to the summing point.

To obtain a sine output, use the two schottky diodes back to back on the summing point to round the triangle; the resulting sine is accurate to about 1% THD.

Adjusting VR_2 gives a symmetrical waveform and you can do it by ear by reducing the frequency so that only harmonics can be heard. The timing capacitor must be changed to 10nF to do this.

With C_1 at 40pF and R_1 at 3kΩ, the *HC4046* works up to 18MHz typical, drawing around 10mA. Values of 10nF and 1MΩ produce 20Hz.

Ian Hegglun

Manawatu Polytechnic
New Zealand

Reference
1. *XR-S200 data sheet, Exar Data Book.*

Speech compressor

An article in the July issue reported a method (Simitar) of increasing loudness of communications signals by up to 20dB, mentioning work at Swansea University about 14 years ago. It is possible that an IC will eventually emerge to perform the function but, until then, the following description should allow people to develop their own system.

Figure 1 shows the principle. Signal is filtered, digitised and sampled by a microprocessor at a rate of at least 8kHz, samples being processed by the micro and output via a D-to-A converter.

At the output, two D-to-As are needed, as seen in

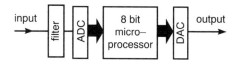

Figure 1. Speech compressor in essence. A sample rate of at least 8kHz is needed and the microprocessor can be any 8-bit variety with simple software.

Figure 2. Input signal from the micro goes to DAC1, the scale factor being written to DAC2 which, in this configuration, act as a divider to modify the output bythe scale factor input from the micro. Any 8-bit microprocessor will work, since no multiplication or division is necessary in the software, although it would be of advantage to use a 16-bit device to do the scaling of signals internally.

Software is shown in flowchart form in the table. The micro stores samples as sign/magnitude values in a 64-wide array (SAMPLES) indexed by pointer (PTR). Before doing that, it takes the old sample and writes it to DAC1, so that the array behaves as a 64-character circular buffer which merely delays the samples. This delay allows the micro to determine the signal peak value between zero crossings, each

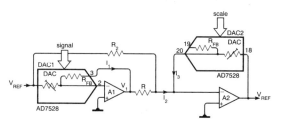

Figure 2. Two separate DACs perform the functions of conversion and scaling.

being compared with the last one (IN0) and with a threshold value (THRESHOLD). If greater, the sample is stored in the array PEAKS, indexed by PIN. When a zero crossing is detected, PIN is incremented and PEAKS and IN0 reset. When samples are retrieved they are also zero-crossing tested, the peak value being taken from PEAKS with the second pointer POUT.

The DACs give the function OUT1/SCALE to scale each half cycle to the same magnitude, which is said to provide a 20dB increase in loudness, subjectively, with none of the usual speech compression problems.

Looking at Figure 2, The DACs are multiplying current-output types such as the *AD7528* from Analog Devices. Current I_1 from the first is

$I_1 = V_{ref}(sig/256)/R_{fb1}$

A1 converts this to

$V_1 = -V_{ref}(sig/256)$,

V_{ref} and the overall feedback resistors converting this offset binary to a bipolar signal

$I_2 = -V_{ref}/R(sig-128)/256$.

Converter DAC2 injects I_3, which is

$I_3 = V_{out}(scale/256)/R_{fb2}$.

Since $I_2 + I_3 = 0$, the output is

$V_{out} = V_{ref} . R_{fb2}/R(sig-128)/scale$.

A somewhat simpler circuit in Figure 3 eliminates the amplifier and two resistors, the gain being R_{fb2}/R_{fb1}, but two low-impedance references are now needed, since high-impedance ones introduce distortion.

David Gibson

Leeds
West Yorkshire

Table 1 (opposite). Software flowchart for implementing David Gibson's processor proposal. There is no connection with Dr Louis Thomas' work.

SAMPLES and PEAKS are 64-wide arrays, indexed into by PTR, PIN and POUT, all initialised to zero.

THRESHOLD holds the lower threshold, below which signal compression does not occur. This stops low level noise from being boosted.

IN1, IN0, OUT1, OUT0 are samples. All initialised to zero.

1.	Get input sample	Read sample into IN1. Format is offset binary.
2.	Get output sample	Take the output sample from the samples array. OUT1 = SAMPLES(PTR)
3.	Store input sample	...and store the input sample in its place. SAMPLES(PTR) = IN1
4.	Increment Pointer	PTR = PTR+1 MOD 64. Thus SAMPLES acts as a circular first-in-last-out buffer of 64 values.
5.	Convert format	XOR $80. Format is now twos complement.
6.	Remove max negative	If IN1 = $80 then IN1 = $81. Maximum negative causes a false zero when next step is performed.
7.	Convert format	If IN1 < 0 then COMplement and add $80. The format is now SIGN and MAGNITUDE.
8.	Test input for zero crossing	TEST (IN1 XOR IN0). IF < 0 then (9) else (11).
9.	Increment pointer	Signs of IN1 and IN0 were different so there was a zero-crossing. PIN = PIN+1 MOD 64
10.	Initialize Peak	PEAKS(PIN) = $7F. Says this signal was of maximum amplitude, so no scaling required. IN0 = 0. Prepares for new half-cycle of comparisons.
	GOTO 12.	
11.	Compare magnitudes	Signs were the same, so compare magnitudes and store the greater one, without its sign. If IN1 U> IN0 AND (S1 AND $7F) > THRESHOLD then PEAKS(PIN) = S1 AND $7F.
12.	Test output for zero crossing	TEST (OUT1 XOR OUT0). IF < 0 then (13) else (15).
13.	Get new scale factor	SCALE = PEAKS(POUT)
14.	Increment Pointer	POUT = POUT+1 MOD 64
15.	Write to DACs	DAC1 = OUT1 (signal) : DAC2 = SCALE
16.	Generate old samples	IN0 = IN1 : OUT0 = OUT1
	GOTO 1.	(Repeat forever)

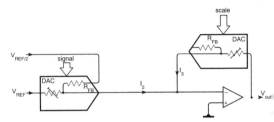

Figure 3. Simpler embodiment of the idea with the slightdrawback that two high-Z voltage references are needed.

Better triggering for oscilloscopes

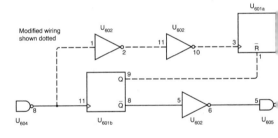

Goldstar OS7020 oscilloscopes exhibit a slight limitation in their triggering facility which causes the sweep generator to trigger only on alternate leading edges (both positive and negative-going) of the signal. Short pulses of low repetition frequency may therefore be difficult to examine in some cases.

This circuit avoids the problem and needs no extra components. Simply cut the tracks close to pins 1 and 2 of U601, the 74LS74, remove the earth link to pin 11 of U602, the 74LS14 near C_{622}, and rewire as shown using the two spare inverters. These are used to delay the clock signal to the blanking and sweep flip-flop, its reset being re-routed to ensure that it is removed before the next trigger at the clock input.

Any additional delay caused by the inverters is negligible compared with that due to the trigger and horizontal amplifiers.

H. Maidment

Wilton

Frequency doubler

One-third of a 74HC14 is used to double the frequency of an input square wave.

The two differentiators $C_{1,2}$ and $R_{1,2}$ convert the input to narrow spikes, only the negative-going ones being selected by the diodes. Since the inputs to the differentiators are in antiphase, a negative-going pulse is present across C_3 whenever the input changes state. At the output of the second inverter, the positive pulse lasts for a time determined by the time constant of C_3R_3, which is adjustable.

Yongping Xia

West Virginia University
Morgantown WV
USA

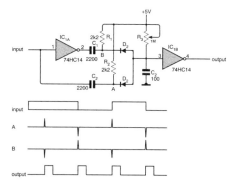

Voltage-tuned crossover filter

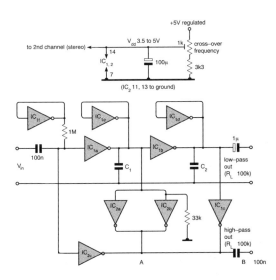

With this circuit, the crossover frequency is variable to suit any loudspeaker, f_c varying over a 5:1 range for a supply voltage change of 3.5-5V. High-pass and low-pass outputs are always in phase.

Since the g_m of unbuffered cmos inverters varies with V_{dd}, connecting input and output together forms a resistor that varies from around 500Ω to 5kΩ for a V_{dd} change of 5V to 3V. Loading an inverter output by a cmos resistive element produces an inverter gain of $|A_v| \approx g_m R_L \approx 1$, since $R_L \approx 1/g_m$. In this case, gain varies from 0.95 to 0.99 for a V_{dd} change of 3.5V to 5V. The element also provides a $0.5V_{dd}$ bias for other inverter inputs.

The filter formed by $IC_{1a,e}$ and C_1 has the same phase characteristic as an all-pass filter. Paralleling $IC_{2a,b}$ gives a gain of 2, the output being summed at A with the output of IC_{2c} to give unity gain and 0° to -180° phase shift with frequency. Inverting the low-pass signal in IC_{1c} and summing at B gives the high-pass function which is in phase with the low-pass output.

The 33kΩ resistor increases high-pass attenuation when V_{dd} is less than 4V to -40dB, but could be left out by using a tuning range of 4V to 5.5V to give a 3:1 frequency range.

Ian Hegglun

Manawatu Polytechnic
New Zealand

Multiple outputs from one D-to-A

Four channels of precision, buffered voltage are obtained from the output of a single D-to-A converter, using the circuit shown which uses fewer components than the conventional method; in particular, there is only one expensive converter against the usual four or more.

A differential multiplexer, IC_5, directs the converter output to each of four low-leakage capacitors and the output buffers in IC_6, all these components being inside the feedback loop of the op-amp IC_4 to reduce offset by the closed-loop gain of the op-amp, which is 101.

A small amount of offset at up to 15μV/°C is still in effect due to the presence of op-amp IC_4.

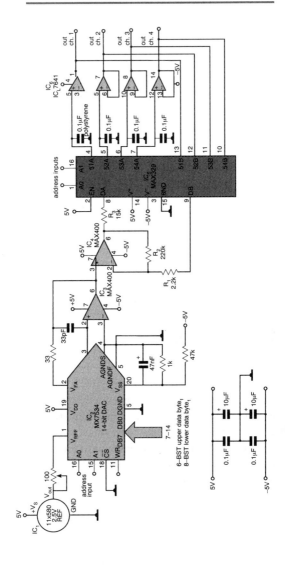

Four channels of buffered, high precision voltage from a single D-to-A converter saves the cost of three expensive components.

Multiplexer IC_5 provides the low current leakage and low charge injection (4pC) needed by the sample-and-hold circuits, the charge injection producing $40\mu V$ offset on each capacitor. When this error is summed with IC_4 offset and that of a buffer amplifier reduced by feedback, total offset is $100\mu V$ or about 0.65 LSB.

With this offset, each hold capacitor takes 2.6s to discharge 1LSB at room temperature; at higher temperatures, this time decreases rapidly and a higher clock rate is needed, although it may mean that several cycles are needed to set up the capacitor voltages after a change.

Jay Scolio

Maxim Integrated Products
Sunnyvale
California

Bi-directional i/o for microcontrollers

I/O pins on microcontrollers which are configurable as either input or output are not usually capable of driving much of a load in one role or providing any schmitt action to shape inputs in the other. For useful interfacing, one must therefore provide drivers or schmitt outputs with, possibly, filters which effectively ruin any chance of software pin assignment.

This circuit does, however, allow more flexibility. The darlington output driver and the schmitt input gate go to the same pin, contentions being handled by software. Values cannot be shown, since they have to be determined by drive capability of the controller concerned and loading on the pin.

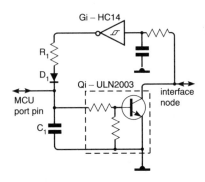

Bi-directional buffer allows configurable i/o pins on microcontrollers to be software controlled without any need for hard wiring.

Input. Set the port pin to "output" and program it low, switching Tr_1 off to prevent it affecting the interface node. Data on the node is inverted by G_1 and it may be found that there is now contention with the port, so R_1 is inserted. To read, the port is momentarily configured as an input. Any tendency for Tr_1 and G_1 to latch disappears when it becomes an output again. If Tr_1 conducts briefly, use C_1. Do not make R_1 too high, or the port pin voltage might not reach 1 because of loading by Tr_1.

Output. When the port is high, the output of Tr_1 goes low, so that the output of G_1 is high and is out of contention with the port. When low, it works against G_1 output for a short time until Tr_1 goes off and G_1 is low; this sets the lower limit of R_1. When the port goes high, it again works against G_1 for a short time; some n-mos controllers have weak pull-ups and the diode assists the action.

David Gibson

Leeds
West Yorkshire

Dial a harmonic

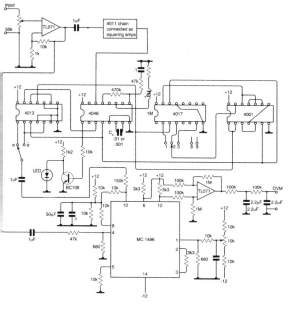

*Dial up your harmonic with this circuit. For low
noise, use precision resistors in critical positions.*

Setting a switch allows the selection of a harmonic of
2nf of an input, for use with a modulator.

At the input, the signal meets a low-signal
amplifier and is then squared in 4011 comparator,
which provides a stabler input and does not need a
multi-turn control. The signal from the amplifier also
goes to the modulator.

Squared signals are the reference in a phase-locked
loop, the n-divider to produce a 2nf frequency
consisting of the 4017, a 4001 and half a 4013,
which divides the 2nf to give in-phase and antiphase

harmonics (nf and -nf). Capacitor C_x determines the range over which the circuit operates and the led indicates lock.

The second TL071 takes the sum and difference frequencies of the two inputs. If the difference is zero, the output is DC, indicating the value of the harmonic n.

You should balance the 1496 modulator by the potentiometer, using an oscilloscope on the TL071 output, with no input (pin 4 shorted).

E Rangel Marins
Instituto Nacional de Tecnologia
Rio
Brazil

Programmable waveform generator

This provides three-function waveform generation from 0.01Hz to 1MHz under the control of a microprocessor, which also indicates frequency digitally.

Control voltage for the 8038 function generator comes from the micro via an 0832 8-bit D-to-A converter, a buffer, a voltage amplifier with a gain of ten and a voltage follower. Sine (top switch position), square and triangular (bottom) waves are taken out through a further voltage follower and variable-gain amplifier, the micro indicating frequency via the 8255 programmable peripheral interface and led display. It is assumed that a linear relationship exists between control voltage and frequency.

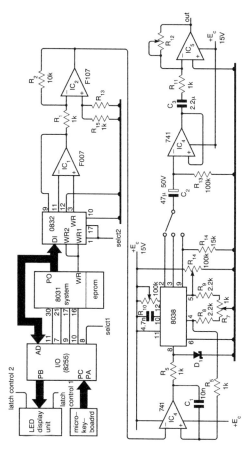

Waveform generator controlled by microprocessor, which also gives a direct indication of frequency.

Sensitivity to frequency control is adjusted by the values of C, R_8 and R_9 and $R_{6,7,10}$ adjust waveform shape. Output voltage is about 20V.

M T Songping Liu

China

UHF VCO

A simple UHF voltage-controlled oscillator for
satellite receivers, offering a frequency range of
1200-2100MHz, output power of +7dBm into 50Ω,
Varicap range 0V-50V and supply requirement of
±12V.

Figure 1 shows a design using inexpensive
components. Parasitic inductance is a major problem
at these frequencies, since it is comparable with
designed inductance in the resonator circuit and
makes decoupling difficult. In this design, some of
the problems are avoided by connecting load to the
collector and the resonator components to base and
emitter.

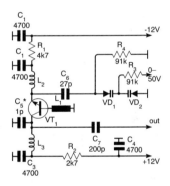

*Figure 1. UHF voltage-controlled oscillator for
satellite broadcast working.*

Construction is much more important in this type
of equipment than that for lower frequencies, so a
description is needed; Figure 2 shows a successful
layout. A 1.5mm double-sided glass-fibre board,
measuring 35 by 30mm is used. Ground pads under

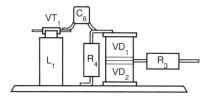

Figure 2. Suggested layout of UHF VCO reduces problems set by stray inductance assuming same proportions as intended variety.

$C_{2,3,5}$ are connected the back plane by plated-through holes. Leads on R_{1-4} an C_{6-7} must be short. It would have been possible to use only one Varicap, but the transistor base would have been taken directly to ground and matching would have been difficult.Capacitors C_{1-4} are disc ceramic; $L_{2,3}$ are 2 or 3 turns of 0.4mm wire on a diameter of 3mm; L_1 is 2.5 by 7mm copper foil, 3mm of it being soldered to ground; the transistor is BFQ69, BFR93, BFR34A or an equivalent, mounted directly onto inductors L_{1-3}.

Raimundas Markevicius
Vilnius
Lithuania

Class-A push-pull amplifier

Mr Richards provides a circuit for an amplifier in which the output bipolar transistors are permanently forward biased. Input signal divides into two halves by means of the diode-feedback "perfect" diodes

IC3,4, while IC2, a simple inverter, is needed to allow overall feedback to be in the correct phase. We have no further information on the circuit.

W O Richards

Battersea London SW11

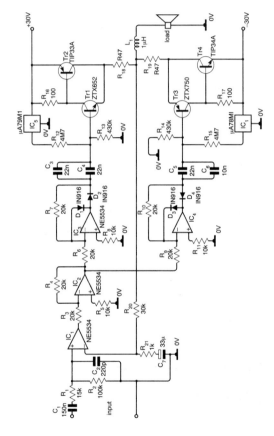

Amplifier has permanently conducting output transistors. Phase splitting is by unity-gain diode-feedback stages and IC$_2$ arranges the various inversions to present overall NFB to IC$_1$.

DC mains inverter

For low voltages and high currents, power mosfets
possess significant advantages over bipolar types.
This 12V input, 240V, 500W inverter is intended to
demonstrate their ease of use and will find
application in power tools, lighting and universal
motors.

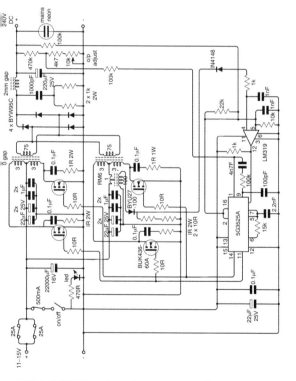

*250V, 500W inverter using power mosfets.
Equivalents to the SGS-Thomson device are
obtainable from Linear Technology, Motorola,
National Semiconductor, Sprague and Unitrode.*

A standard push-pull inverter runs at 25kHz, with regulation obtained from the voltage-mode SG3525A pulse-width modulator from SGS-Thomson. Current-limit shutdown is achieved by sensing the current in the centre tap of one of the transformers (not both, since they are identical), an op-amp comparator driving the shutdown IC input. Threshold of current limit at the comparator is reduced when the inverter input is very low to compensate for delay in shutdown.

On a practical note, the high primary-side currents render a ground plane more or less essential and copper tape primaries definitely essential, although 1mm enamelled copper will suffice for the secondaries.

Paul Bennett
Stoke Gifford
Bristol

A different bridge

This circuit simulates a Wheatstone bridge (Figure 1) by an operational transconductance amplifier and avoids the need for matched resistors while providing a better signal from a strain-gauge transducer. An OTA, as shown in Figure 2, gives a current output $I = G(V_+ - V_-)$, in which G is transconductance, variable by means of R_b, the biasing current control. If G is adjusted to be $G = 1/R$, $Vo = IR = V$, the second amplifier output being zero in this case, simulating a balanced bridge; a small variation in the value of R produces a linear output.

Responses of the two circuits for the same reference voltage and variation in resistance ∂R is

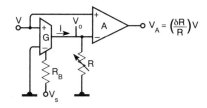

Figure 1. Standard Wheatstone bridge needs matched resistors.

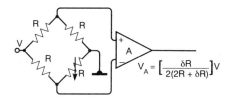

Figure 2. Novel bridge arrangement using operational transconductance amplifier avoids need for matching and provides better output from a strain gauge.

given for comparison. In the prototype, the OTA was a CA3080 and the op-amp an AD524 .

Promit Biswas

New Delhi

AC stabiliser

Originally designed to stabilise the supply to an over-sensitive refrigerator, this arrangement will handle a 220V supply at 20W to 2kW.

Two circuits are offered for resistive and reactive loads, the second being a modification of the first. Figure 1 shows the layout of the resistive-load version. When the mains goes negative with repect to the diode D_4, all the current passes through D_5, control not being effective during this half-cycle. On the positive-going half-cycle, D_4 voltage increases, but no current passes since it is not yet triggered.

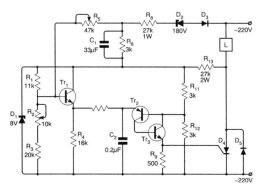

Figure 1. Resistive load version of AC mains stabiliser.

Current through Tr_1 and R_7 charges C_2 to around 5V, at which point the n-p-n/p-n-p pair fires, discharging C_2 and triggering the thyristor by the voltage across R_9. Duration of current cut-off through the thyristor is determined by D_1, $R_{1,2,3}$. Figure 2 is a graph of the relevant currents and voltages.

Components $D_{3,2}$, $R_{8,5,6}$, C_1 provides a supply voltage to the control circuit proportional to the mains voltage. When the mains voltage increases, the supply to Tr_1 decreases, extending to cut-off time of the thyristor and vice versa, average current remaing unchanged. Zener D_2 reduces the effect of the rectifier on the rest of the external circuit.

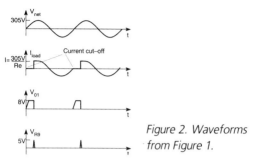

*Figure 2. Waveforms
from Figure 1.*

Figure 3 shows the modification for reactive loads,
which works symmetrically on both half-cycles, its
operation being sketched in Figure 4. Figure 5
details the precautions necessary to avoid affecting
the local mains supply by HF disturbances.

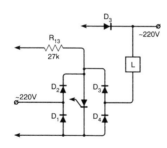

*Figure 3. Modification to thyristor circuit for
reactive loads: thyristor is set in bridge.*

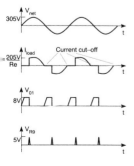

*Figure 4. Waveforms
with the modifications
of Figure 3.*

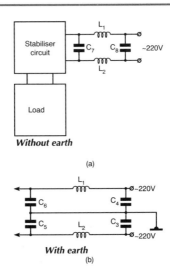

Figure 5. Interference suppression circuits.

Transistors $Tr_{1,3}$ may be BC109s, MPSA06s or similar and Tr_2 a BC212, MPSA56 or the like. The resistor R_{13} should be of 2W rating and R_8 a 1W type.

Valery V Vershinkin

Avdeyevka
USSR

Low-current transducer driver

A standard oscillator driving a three-terminal piezoelectric transducer (Figure 1) may be extended by a few extra components to give the same output while drawing much less current.

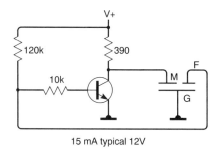

15 mA typical 12V

Figure 1. Standard oscillator for use with three-terminal piezo transducers.

Figure 2 shows the circuit, in which the diode D1 forces the oscillator to run at 50% duty cycle – the optimum ratio. Values are uncritical and the circuit is happy with a supply of 5V to 30V. Beware reducing current demand by increasing the collector resistor too much, since in that condition starting becomes unreliable.

Mark Byrne

Devizes
Wiltshire

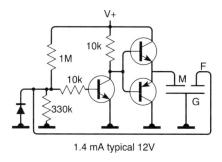

1.4 mA typical 12V

Figure 2. Modification to reduce current drawn by oscillator, while retaining same output.

Programmable pulse train generator

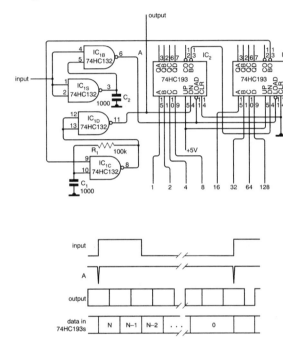

Number of output cycles of oscillator depends on data applied to presettable counter data inputs.

When an input triggers this circuit, the reuired number of pulses from 2 to 256 automatically appear at the output.

Nand sections $IC_{1a,b}$ form the input triggers into very narrow load signals which enter the the input data (the pulse number) into the 4-bit binary counters $IC_{2,3}$. Data being loaded, the B_0 output of IC_3 is 1 and allows an oscillator IC_{1c} to function and to step the two counters down until they are both at zero, at which point the B_0 output is 0 and the oscillator stops, having produced a number of output cycles dependent on the input data plus one count since the counters go to 0. Oscillator components R_1, C_1 determine the output frequency.

Yongping Xia

West Virginia University
Morgantown,
WV, USA

Busy line indicator

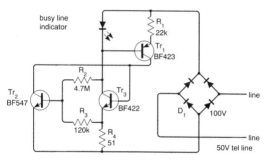

Telephone line activity indicator uses a mere 50μA in standby mode and may be modified for use in control.

Telephone line-activity indicator shown is simply connected in parallel with the line. It is line-powered and takes a stand-by current of less than $50\mu A$, active current being about 8mA.

Line voltage is rectified for ease of installation and is needed anyway if outgoing calls are inverted. Transistor Tr_2 senses the 50V line voltage and turns Tr_3 off in standby mode, sensing voltage across R_4 to limit active-mode current to avoid latching. Base current to Tr3 comes from Tr_1, current being limited by R_1. A 120V V_{CEO} rating is needed by Tr_1 and Tr_3 to avoid breakdown to ringing voltage.

If an opto-coupler is used instead of the led, it will control recording equipment and transmitters etc; a small amount of modulation is visible in the light output, which indicates voice traffic.

Ron Weinstein

Centralab
Tel Aviv
Israel

High-res A-to-D using low-res converters

Using n low-resolution A-to-D converters, increase the final resolution of a converter by n-fold by means of the circuit shown here.

Converter 1 gives the most significant byte of the conversion, of which the analogue equivalent is at the output of the A-to-D converter and is subtracted from the analogue input by the 741 difference amplifier, providing a gain of 256. This voltage goes to the second A-to-D, which produces the least significant byte of the conversion, the end-of-conversion signal from the first serving as the start-conversion input for the second, whose EOC signals the end of the whole conversion. So two 8-bit A-to-D converters function as one 16-bit device.

In principle, resolution of an A-to-D converter increases by n times for n low-resolution A-to-Ds of most kinds, including flash and successive-approximation types.

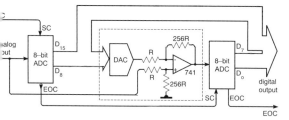

Two low-resolution A-to-D converters perform the function of one high-resolution type, the principle being capable of extension to an increase of n times.

K Balasubramanian

Cukurova University
Adana, Turkey
Turkey

Adjusting differential amplifier gain

Figure 1 shows the common-or-garden differential amplifier, which is known to be simple and reliable – unless its gain is to be made variable. In that case, ganged potentiometers or another gain stage might be needed, which neatly remove the advantages of simplicity and reliability.

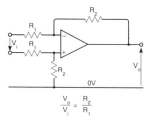

$$\frac{V_o}{V_i} = \frac{R_2}{R_1}$$

Figure 1. Ordinary differential amp for fixed-gain operation.

Figure 2 is one way out; if R_g is large compared with R_1, gain is $\approx R_2/R_1$, whereas a small R_g gives a gain approaching zero. In the case of o, the reverse applies; a small R_g gives a high gain, a large value confers a gain of R_2/R_1. You cannot carry this too far, however, since if you make R_g too small, negative feedback is no longer effective.

B D Runagle

Swadlincote
Derbyshire

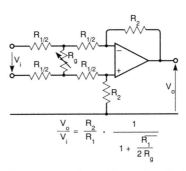

$$\frac{V_o}{V_i} = \frac{R_2}{R_1} \cdot \frac{1}{1 + \dfrac{R_1}{2\,R_g}}$$

Figure 2. One way of providing gain adjustment without losing benefits of simplicity and accuracy. Gain is from zero to R_2/R_1.

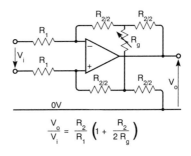

$$\frac{V_o}{V_i} = \frac{R_2}{R_1}\left(1 + \frac{R_2}{2\,R_g}\right)$$

Figure 3. Another way, giving adjustment from R_2/R_1 to a high value.

Simple, but versatile timer

One non-inverting cmos buffer and three passives compose this simple timer.

Point A is normally low, whereupon the output B is also low. When A goes momentarily high, C_1 charges through D_1 and the output goes high, where it stays for a period of up to about five minutes when A is taken low again, the exact time being determined by the CR.

You could make the circuit a little more flexible by replacing R_1 by a 2.7MΩ potentiometer having a 10kΩ fixed resistor in series to give variable times.

K R Kirwan

Aldersley
Wolverhampton

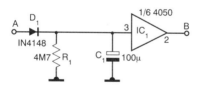

Simple timer for long time intervals, which could be made variable.

Continuous on/off timer switch

On and off times of this continually operating switch are settable from seconds to hours independently of each other.

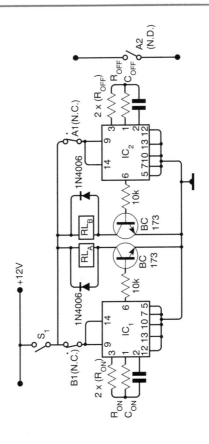

Timer produces on and off periods, independently adjustable from a few seconds to hours.

Closing S_1 applies power to IC_1, an MC14541B oscillator/timer, making the associated BC173 conduct and energise RL_A. Contact A_1 opens to de-energise IC_2 and close contact A_2 – the load switch. After IC_1 timesout, A_2 opens to isolate the load and A_1 closes, applying power to IC_2. Relay RL_B is now energised and contact B_1 opens, disconnecting IC_1. When IC_2 times out, IC_1 is once again under way and the whole thing starts again. Values of R_{on}, C_{on}, R_{off}, C_{off} are given by $T = 1.15R_C \times 8192$.

John Karageorgakis

Thessaloniki Greece

Reference
Motorola data sheet on MC14541B programmable timer.

Divide by 2.5

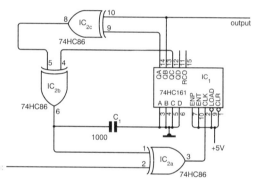

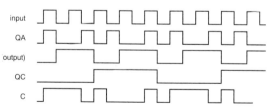

Circuit by Xia to divide input frequency by 2.5. Circuit can be made to trigger from either edge.

Division by 2.5 is performed by two ICs, one of them being a 74HC161 positive-edge-triggered 4-bit binary counter.

A logic level at point C determines whether the counter triggers on a positive-going or negative-going edge at the trigger input terminal, since trigger input is fed via an X-Or gate IC_2, C=1 giving triggering on a negative-going edge. Level at C is controlled by the counter outputs and the circuit output from QB is a cl ean, though asymmetric waveform.

Yongping Xia

West Virginia University
Morgantown, WV
USA

Dual-speed DC motor controller

A cmos NAND IC, the CD4011, is the core of a pulse-width controller for DC motors, providing logic selection of two preset speeds.

Gate G_4 forms one half of two separate astable multivibrators activated by a logic signal to G_1. One of the astables is formed by G_4 and the components C_1, preset P_1 and diodes D_1 and D_2, the operative NAND being G_2; preset P_1 sets the mark:space ratio for this astable.

When the speed selection input is low, this astable oscillates under the control of P_1 and drives the output transistor; when high, the otherastable takes over at a M:S ratio set by P_2. Run and Stop control is a separate input.

M S Nagaraj

ISRO Satellite Centre
Bangalore
India

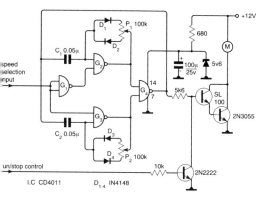

*Two-speed pulse-width control of a DC motor
with logic input; the two speeds are independently
set.*

Millihertz multivibrator

When accurate and repeatable long-period timing is
needed, a CD4053 analogue multiplexer and a
CD4020 14-stage counter will form an astable
multivibrator with independent setting of mark and
space from seconds to minutes. No expensive large-
value, low-leakage capacitors are needed and there is
no initial pulse error, often present in analogue
timing circuits.

Analogue switches B and C of the 4053 are
arranged as inverter gates, inputs and outputs
therefore being complementary. Capacitor C_1 and
resistors $R_{A,B,C}$ form an oscillator round the gates.

Pulses from the oscillator are counted by the 4020,
its Q_{14} output changing state after every 8192 pulses
at the input and controlling, by way of gate A, which
of the two timing resistors R_A or R_B is in the
oscillator circuit. The frequency of the oscillator is so
decided by the value of R_A when Q_{14} is low and by

R_B when it is high, Q_{14} being taken as the output.

With the component values shown in the diagram, on and off times are variable between 10s and 70min. Leakage in the off channel of gate A causes negligible change in the timing of each period when the timing resistor of the other period is varied.

If the preset feature of the 4020 is not needed, a 4045 21-stage counter can be used to give longer periods.

M S Nagaraj

ISRO Satellite Centre
Bangalore India

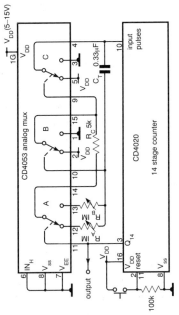

Low-power counter and multiplexer give independent control of on and off periods of multivibrator from 10s to 70min, using no expensive capacitors. Timing is much more accurate and repeatable than in analogue timers.

RIAA amplifier uses current feedback op-amp

In a current feedback op-amp such as the Analog Devices AD844 with access to the buffer, the active devices to make an RIAA amplifier are in one package.

An RIAA network on the input to the buffer, which is a unity-gain stage, effects the required equalisation. Input impedance is equal to R_{in}, since the inverting input is at low impedance without the assistance of external feedback, and DC gain is $A_v = R_1/R_{in}$. Slew-rate limitation is not a problem with the AD844, rise and fall times of the output being almost independent of the amplitude.

A point to watch is that maximum current through R_1 is around 10mA, which must be borne in mind when setting gain and overload margin. Decoupling components should be close to the pins.

Richard Heycock

Beverley
North Humberside

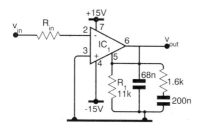

AD844 transconductance amplifier/unity-gain buffer combination enables this simple realisation of an RIAA amplifier.

85mA op-amp output

Output current from op-amps is commonly between 20 and 40mA. In this LM324 quad op-amp circuit, a parallel connection of three amplifiers gives up to 85mA and retains short-circuit protection.

This is a unity-gain circuit providing a bandwidth of 200kHz, R_1 and C_1 taking care of loop stability. Current-sharing resistors of 10Ω avoid the effect of inherent differing input offsets, which would otherwise cause different output currents from each op-amp. For long-term, high output current, a bonded heat sink will probably be needed.

A M Wilkes

Brentwood Essex

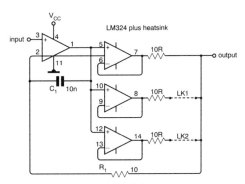

This parallel connection of the op-amps in one package will give up to 85mA output current. Bandwidth of this circuit is 200kHz at unity gain.

Measuring transfer functions

In a frequency-swept transfer-function analyser for use with an oscilloscope, it is necessary to maintain the level of the swept oscillator constant at all relevant frequencies – a problem that often exercises designers. In my design, the problem is largely eliminated

Since a square wave is composed of an infinite series of harmonics, the fundamental being predominant, the LF response of a circuit to a square wave indicates its response to the fundamental. As Figure 1 shows, if A is fed with a square wave, its LF response V_{OL}, which is the flat part, accounts for the transfer function at that frequency. If, therefore, a series of period-modulated square waves is applied to a circuit, the relative amplitudes of the flat parts

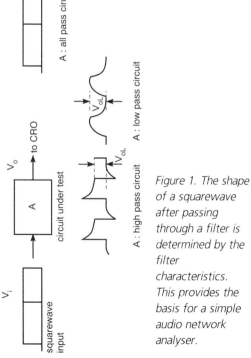

Figure 1. The shape of a squarewave after passing through a filter is determined by the filter characteristics. This provides the basis for a simple audio network analyser.

of the response gives the frequency response to all the fundamental frequencies of the input series.

This the core of my transfer-function scanner. It produces a series of square waves that are frequency-modulated in steps, rather than continuously. Figure 2 shows the essentials. A 555 timer provides the clock, which drives the 4017B decade counter. This, via the *4066B* analogue switches, connects C_{1-n} in the second 555 timer (Z). The output is therefore a number of stepped frequency changes, the step duration being fixed by the clock and the square-wave frequency in each step by R_1C. Figure 3 is a more detailed diagram, in which values may be chosen for any particular application. In the circuit described, the frequencies inside the steps varied in a 1-2-4 sequence from

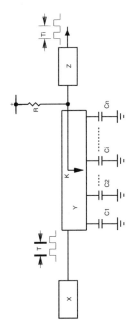

Figure 2. Essentials of the transfer-function scanner.

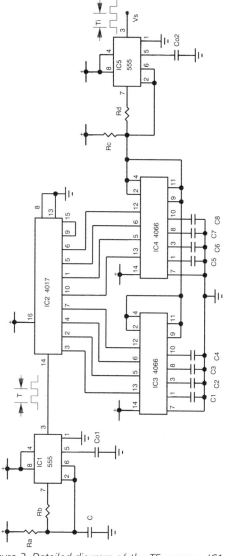

Figure 3. Detailed diagram of the TF scanner. IC1,5 are 555s, IC2 a 4017B and IC3,4 4066Bs, although the chip count could be reduced by using a 556 and a 4067.

100Hz to 12kHz and the capacitors C_{1-8} from 100nF to 800pF. Resistor R_c is 72k and R_d is 30.5k.

Figure 4 shows some results. At (a) is a screen shot of a high-pass filter circuit, showing a clear

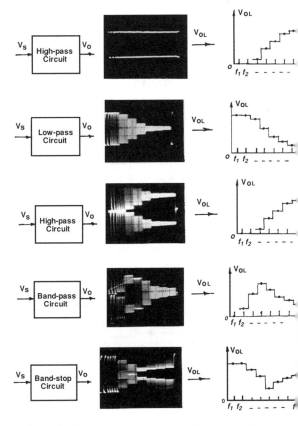

Figure 4. The results from various filter types. The circuit in Figure 3 produces a step-frequency mdoulated squarewave output which, when applied to a filter input, produces output voltages and waveforms of the types shown below. The swept frequency oscillogram requires a degree of interpretation – the RMS voltage output does not correspond to external envelope shape.

representation of the effect, with the measured response on the right. At (b) is a low-pass response and the output of the transfer-function scanner itself at (c).

T C Liao

Beijing Peoples's Republic of China

DS1233 replaces monostable

Dallas's *DS1233 EconoReset*, described in the March 1992 issue (p910) normally resets a microprocessor after detecting upsets on its supply, but has other possibilities, being effectively a monostable in a TO-92 package which maintains a low on the output for 350ms after power is applied. Here, it delays and produces an inverted pulse with a fixed width. Its frugal power needs can be supplied by a cmos gate.

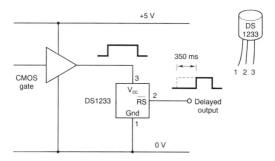

Figure 1. Dallas's Econoreset microprocessor reset device used to delay a pulse rising edge by a fixed 350ms. Used in such a way, the DS1233 replaces a monostable and an And gate is contained in a small package. Power comes from the cmos gate.

Figure 1 shows the former, in which it delays a rising edge by 350ms, replacing a monostable and an and-gate. When the cmos gate output rises, the *DS1233* output remains low for 350ms, going high after that period and returning low when power is removed.

In Figure 2, a negative-going pulse wider than 350ms at the gate output produces a 350ms pulse from the *DS1233*. When the gate output goes low it applies power to the *DS1233*, the output going low for the time-out period and then returning high, unless the gate output is shorter than 350ms, in which case the output will correspond to the input.

Steve Winder

Ipswich, Suffolk

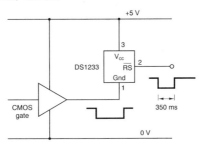

Figure 2. The DS1233 produces a fixed 350ms negative-going pulse.

Variable-inductance, low-frequency VCO

Variable-frequency oscillators using the then principle of varying inductance of a coil by varying mutual inductance in a transformer were first described by K C Johnson in *WW* April and May 1949. My adaptation is shown in the diagram.

If L is the inductance of a coil through which flows an alternating current and some part of the same current flows in a mutually coupled coil, the effective inductance of the first coil L_e is $L_2 \pm M$, since the second coil may or may not be wound in the same sense. M is the mutual inductance.

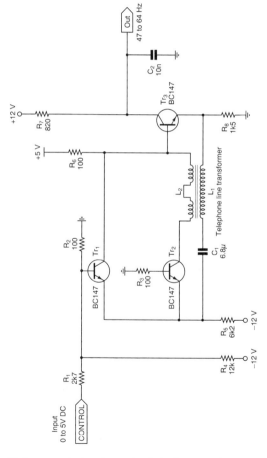

Voltage control varies series inductance and therefore frequency in this LF oscillator.

A differential voltage-controlled amplifier can be used to vary the proportion of the oscillatory current flowing through the second coil, the total oscillatory emitter current being shared in a varying proportion between the two halves of the amplifier. Since the effective inductance is in a series resonant circuit, the oscillator frequency also varies.

Transistor Tr_3 is an emitter follower feeding the common-base amplifier made up of Tr_1 and Tr_2, the differential pair, whose output goes to Tr_3 and completes the loop. Loop gain is set by $R_{68\Delta 2}$

to just over unity. Frequency is determined by C_1 and L_e, although since the coil in Tr_2 collector passes a direct current, ferrite cores affect the frequency.

In the oscillator shown, L_1 and L_2 were made from a telephone exchange line transformer, which gave a frequency range of 47-64Hz for a 0-5V input voltage. Squegging at 600kHz occurred at zero crossing points, which was eliminated by the addition of C_2; a more suitable transformer may be designed to avoid the problem.

Mike Button

TDR Ltd,
Malmesbury
Wiltshire

Under-frequency inverter protection

If a 50Hz inverter's output frequency falls below that required by the equipment it powers, this circuit disconnects the output.

Input comes from the inverter's driver stage, any asymmetry being eliminated by the first D-type flip-flop. At each low-to-high transition of Q_1, C_T

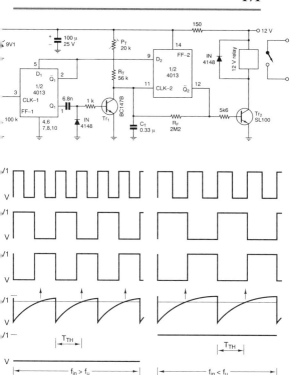

Simple circuit disconnects inverter output if its frequency falls below a preset limit. It could easily be used with portable AC generators.

discharges through the transistor Tr_1 and begins to charge again through R_T and P_T. As this voltage reaches the threshold voltage of Clock 2 input, the $\backslash Q_1$ output latches into the second flip-flop. As shown in the timing diagram, the $\backslash Q_2$ output, which drives the output transistor and therefore the relay, is either 0 or 1, depending on the input frequency. Resistor R_F inserts a little hysteresis to prevent relay chatter.

Adjust P_1 to make $T_{TH} = T_u$ ($=1/f_u$, the frequency at which the relay disconnects the load). This trip frequency can lie in the 48-62Hz range with the components shown.

M S Nagaraj

Isro Satellite Centre
Bangalore India

Single-diode full-wave rectifier

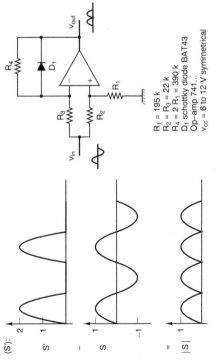

Using only one diode and a single op-amp, this is a low-frequency, full-wave rectifier

This single-diode, single op-amp rectifier is used for LF rectification in an RTTY FSK demodulator.

During a positive input half wave, D_1 conducts and the circuit becomes a non-inverting amplifier, so that

$$V_{out} = V_{in} \frac{R_1}{R_1 + R_2}$$

for positive inputs. On negative half cycles, D_1 is virtually an open circuit and
on negative inputs.

$$V_{out} = V_{in} \frac{R_1}{R_1 + R_2} \left(1 - \frac{R_2 R_4}{R_1 R_3}\right)$$

Making $R_3 = R_4$ and $R_1 = 2R_2$ reduces this to

$$V_{out} = V_{in} \frac{-R_1}{R_1 + R_2}$$

which is the inverse of that for positive inputs and both half cycles are amplified.

Diode imperfections can cause an imbalance between the halves; in such cases, increase R_1 to 200-220kΩ.

Francois Guillet
France

Cascode oscillator

This is a simple oscillator, but is very reliable and exhibits many of the features of a near-perfect circuit.
It uses a two-terminal coil with no taps, self starts and draws only 1mA at 12V, the drive being inherently Class D. Output impedance is small and output swing large. Over the supply-voltage range of 2-24V, frequency stability and waveform purity are

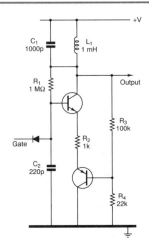

Reliable and frugal oscillator has low output impedance and is easily gated to start consistently at the same point.

exceptional. Taking the gate input to 0V starts the oscillator at the same point in the cycle and the earthy end of R_4 has a similar, but reversed effect. Tuned circuit L_1C_2 determines frequency – 160kHz in the case shown – and the time constant R_1C_2 must be longer than the period.

J J Hyland

Glazertron Ltd
Rochester Kent

Battery backup

This provides 12V DC to an alarm clock when the mains-derived supply fails, the 12V coming from four 1.2V cells in a boost converter. The circuit is simply connected in parallel with the existing DC supply, which maintains charge on the cells through R.

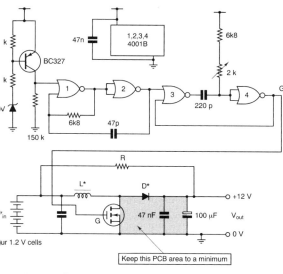

DC-to-DC boost converter backs up 12V DC supply, using off-the-shelf components. Cells are normally on charge.

If the mains supply falls, the *BC327* circuitry detects the drop and turns on the 1MHz multivibrator of gates 1 and 2, which triggers the monostable of gates 3 and 4, the output pulse of which is variable in width.

The pulse, G, drives the converter mosfet, a low-voltage, low-$R_{DS(on)}$ type such as the *BUZ10*. Diode D is an ultra-fast device (an MUR110 was used) and inductor L is five turns of doubled 0.2mm diameter enamelled-copper wire on a ferrite bead, inductance being 20μH. It must not enter saturation.

With a 100kΩ load, 3V from the cells gives 12V output, this increasing to 5V for loads of 1kΩ or less down to a practical limit of a 220Ω load (around 50mA).

Dominique Bergogne

Saint Etienne France

Electrolytic ESR tester

In sensitive circuitry, for example in a feedback loop, it is often necessary to know the equivalent series resistance of an electrolytic capacitor. This circuit measures ESR quickly and simply, assuming access to a digital storage oscilloscope.

Operation is simple: press the push-button switch and view the DSO trace. Calculate ESR from $ESR=(v_1/v_2)-1$ in ohms, the two voltages being those indicated in Figure 2.

Replacement of the pushbutton switch with a logic switched mosfet would eliminate switch bounce effects. It would also allow operation at higher voltages for greater signal output. The channel resistance and self capacitance of the device need to be taken into account however.

A M Wilkes

Glasgow

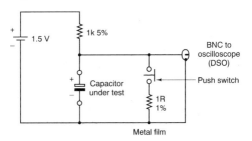

Figure 1. Very simple circuit to measure equivalent series resistance of an electrolytic capacitor.

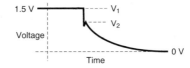

Figure 2. Trace on DSO resulting from capacitor discharge.

Four channels on a single-channel oscilloscope

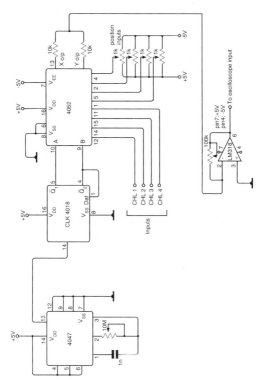

Circuit for multiplexing four y channels into one single-channel oscilloscope input.

On a single-channel oscilloscope, this circuit, using only four ICs, multiplexes four signal channels for, effectively, simultaneous display.

The differential 4052 multiplexer works with two sets of four inputs: pins 11,12,14 and 15 carry the y signal, while pins 1, 2, 4 and 5 take DC potentials from the four potentiometers to determine the y position on the screen (a DC-coupled oscilloscope is assumed).

Clock pulses variable up to 2MHz are generated by the 4047 astable and drive the 2-bit Johnson counter, which produces A and B select waveforms for the multiplexer. High switching rates multiplex the y inputs at a higher rate than the oscilloscope sweep to give a virtually continuous display.

Output to the single input of the oscilloscope comes from an LM318 variable-gain op-amp.

V Lakshminarayanan

Centre for Development of Telematics
Bangalore, India

Accurate astable multivibrator timing

In the free-running multivibrator of Figure 1 the period is theoretically given by $T = 2CR_1 \ln(1+2R_2/R_3)$, positive and negative excursions being exactly equal.

Symmetry of the output waveform suffers in practice when the differential input voltage exceeds the specified value and, in some types of op-amp,

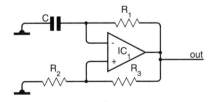

Figure 1. Commonly used multivibrator circuit may give asymmetrical output if inverting input avalanches.

causes an avalanche input current. This usually happens as a result of exceeding input threshold protection or when long-tailed pair base/emitter zener action occurs.

Figure 2 shows an improved circuit, in which the unity-gain buffer IC_2 isolates the timing circuit from the inverting input of IC_1, incidentally affording

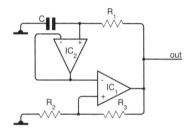

Figure 2. Modified circuit isolates timing components from inverting input to preserve symmetry.

increased speed. Alternatively, the potential divider $R_{2,3}$ may be varied to give a lower differential input swing, but in this case the op-amp offset voltage cannot be neglected.

Dmitri Danyuk and George Pilko

Kiev , USSR

Variable M:S op-amp oscillator

Needing a low-frequency oscillator to give a rough idea of the level of a control voltage, I considered the obvious 555 but, having a spare op-amp, developed this variation of a common design.

Op-amp oscillators commonly take the form shown in Figure 1, in which R_1 and R_2 provide positive feedback and the reference voltage. Capacitor C charges through R_3 to the reference level, whereupon the output goes low and C then discharges to the new level. When C reaches it, the output again goes high and the cycle repeats. With equal supplies, the mark:space ratio is 50:50.

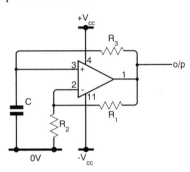

Figure 1. Common form of op-amp oscillator, providing 50:50 mark:space ratio with equal supplies. Feedback resistors determine reference voltage.

In Figure 2, a single supply is used and the low impedance of the control voltage source enables it to be used to vary charge and discharge times by changing the switching point of the op-amp to alter the M:S ratio.

Values given provide a led flashing rate of 3Hz, assuming a low-leakage C, and M:S ratio is variable from 3:1 to 1:3 for a control voltage swing of +2V to +10V. I used a low-current led to 0V, but a higher-current device could be driven by a transistor.

Martin J Barratt

Reading, Berkshire

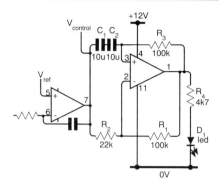

Figure 2. New circuit with single supply. Control voltage to be monitored varies mark:space ratio of waveform driving led indicator.

Simple x2 converter

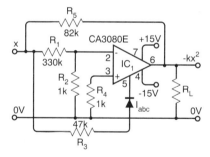

Simple circuit gives kx^2 where k is about 0.1, and avoids need for four-quadrant multiplier.

To avoid the complication and expense of using a four-quadrant multiplier to produce an input-squared output, I have used the circuit shown.

An operational transconductance amplifier computes the expression $-kx(x+v)$, where V is the negative supply voltage and k depends on component values. A term ($-kxV$) is now present at the output, proportional to the input, which must be removed by adding an equal but opposite component through R_5. The result is $-kx^2$, where

$$k = (g_m R_2 R_5 R_L)/[(I_{abc} R_3 (R_1 + R_2)(R_5 + R_L)]$$

In the circuit shown, $k \approx 0.1$ if R_L is high. An offset null adjustment and gain compensation provide greater accuracy.

Ian M Wiles

Basingstoke
Hampshire

Parallel-to-RS232-C conversion

The circuit shown in Figure 1 converts parallel 8-bit data at the inputs of the uart IC_2 (pins 26-33) to RS232-C serial data format for transmission at any of fourteen of the commonly used transmission rates as in Table 1. Data is entered either by means of switches DWS1-DWS8 or from an external source.

Transmission is initiated by applying a positive or negative edge to the trigger signal input of the multiple monostable IC_3, an HEF4528; selection of trigger polarity is by switches TS 1-4. Serial data from pin 25 of the IM6402 uart is inverted and converted to RS232-C levels by the Max232 driver, IC_4, appearing on pin 14. The uart inserts a start bit and either one or two stop bits and odd, even or no parity bits, selected by switches CW 1-5 as in Table 2. Figure 2(a) shows the data format and Figure 2(b) the timing.

If data is to come from an external source (for example, from the motion detector published in EW

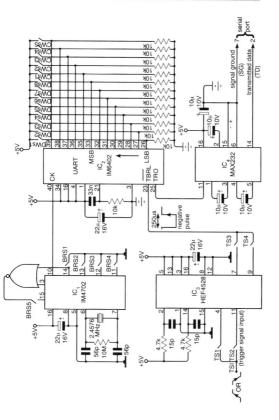

Figure 1. Parallel-to-RS232-C converter for external data or for switched-input setting.

+ WW, July, pp. 571, Circuit Ideas), the switch/resistor network on pins 26-33 of the uart is omitted. In this case, valid data must be present at least 50ns before and 70ns after the trailing edge of the negative pulse at pin 23 of the uart.

K Kumaran

University of Keele

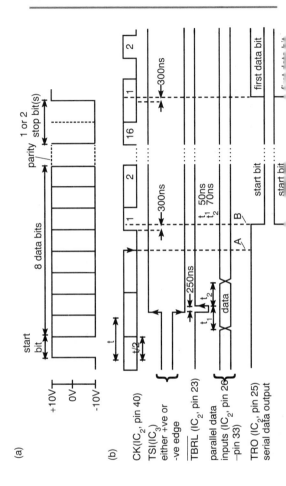

Figure 2. Serial data format and levels at output of RS232-C driver (a). Number of stop bits is optional, as is inclusion of odd or even parity bit. Timing of circuit is at (b). Point A is first negative edge of clock, at least T/2+175ns after positive edge of TBRL; B is start bit time, T/2+300ns after A.

Rate	BRSl	BRS2	BRS3	BRS4	BRS5
50	on	off	on	on	off
75	off	off	on	on	off
110	off	off	off	off	off
134.5	on	on	off	on	off
150	on	off	off	off	off
200	off	on	off	on	off
300	off	on	off	off	off
600	on	off	off	on	off
1200	off	off	on	off	off
1800	on	off	on	off	off
2400	on	on	off	off	off
4800	off	on	on	off	off
9600	on	on	on	off	off
19200	off	off	off	off	on

Table 1. Selecting data rate by setting of switches BRS 1-5.

CWS 1	CWS 2	CWS 3	CWS 4	CWS 5	Parity	Stop bit(s)
on	off	on	on	x	none	1
on	on	on	on	x	none	2
off	off	on	on	off	odd	1
off	on	on	on	off	odd	2
off	off	on	on	on	even	1
off	on	on	on	on	even	2

x = don't care

Table 2. Selection of start, stop and parity bits.

Tolerance-independent D-to-A

If conversion time is not important, this circuit avoids the use of high-precision resistors while still providing extremely high linearity.

In the diagram, the outputs of an 8-bit counter are compared with an 8-bit digital input in the LS684 magnitude comparator. The resulting rectangular

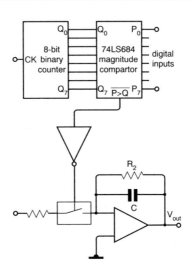

Digital-to-analogue converter does not depend on resistor tolerances.

wave at the $P>Q$\ output has a mark:space ratio directly proportional to the P input and is used, via an inverter, to turn on and off the analogue switch. Direct current through R_1 and R_2 is therefore modulated by the digital input and the amplifier output is $-(NV_{ref}/256)(R_2/R_1)$, where N is the nemerical value of the digital input. Resistor values have no effect on linearity, merely setting conversion factor. Capacitor C smoothes the output.

Linearity depends in the main upon amplifier performance, assuming the frequency is not too high. It should be a simple matter to modify the counter and comparator to provide more bits of resolution.

David J Haigh

Stoke

Coventry

D-to-A converter on Centronics printer port

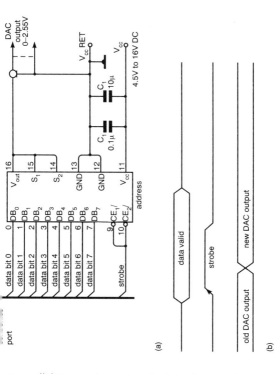

A parallel Centronics port and a D-to-A converter produces a voltage between 0 and 2.5V for a number input to the port of 0-255.

Very few external components in addition to an 8-bit A-to-D converter will give an analogue voltage proportional to the data on a parallel Centronics printer port.

Depending on the number between 0 and 255 at the port, the voltage level lies between zero and

2.55V. Port data is stored in the *AD558*'s internal latch as the strobe goes high, although the strobe may be inverted if necessary to take account of those machines in which data is not valid on the rising edge of the strobe.

J Vandana

World Friends AI Group
Tamilnadu
India

Pulse-width monitor

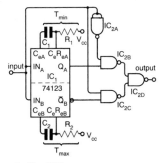

note : IC_2 = 7400
note : pins $\bar{Q}_A$ and Q_B of 74123 are unused

Positive-going pulse at the output signals an input pulse outside a preset time range.

If a positive-going pulse is either shorter or longer than a preset time, this circuit indicates the fact.

Minimum time T_{min} and maximum T_{max} are adjusted by means of C_1R_1 and C_2R_2, the timing circuits of a dual monostable flip-flop. Outputs of the two Nands IC_{2b} and IC_{2c} remain high unless an out-of-range pulse is detected, in which case one goes low and the output of the circuit produces a positive-going pulse.

Frantisek Michele

Brno
Czechoslovakia

Steep-cut, low-pass filter

In Circuit Ideas for July 1987, Tim Mason showed a filter circuit which had the requirements of a steep roll-off above 15kHz with a deep notch at 19kHz. A computer model of the filter gives the result shown in Figure 1, where the peak is caused by the undamped parallel-T section.

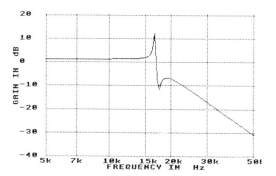

Figure 1. Original steep-cut low-pass filter with notch at 19kHz.

In my modified version shown in Figure 2, the parallel-T gives low-pass filtering as well as the notch by attenuating the input to the high-pass section; resistor R_2 dampens the peak by reducing positive feedback. Notch frequency is 19kHz.

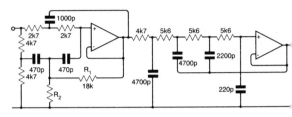

Figure 2. New circuit eliminates peak at 16.5kHz, gives deeper notch and a greater degree of low-pass filtering.

To obtain the fairly sharp corner in the response at around 15kHz, 2% components are necessary. A source impedance of <200Ω, such as that from an op-amp, is also needed. Amplifiers in the *TL071/2/3* series, with an open-loop unity gain frequency of 1MHz or better should be used.

Scaling capacitor values up by a factor of 10 would give a corner frequency of 1.5kHz — useful for speech on narrow-band SSB direct-conversion receivers; scaling by 100 would give 150Hz for Morse.

J A H Edwards

Leicester

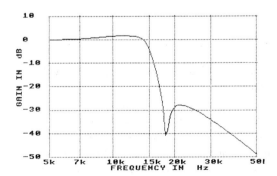

Figure 3. Response of Edwards circuit.

Rumble filter preserves bass

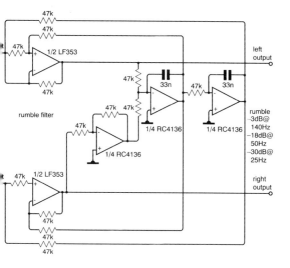

Out-of-phase signals caused by turntable rumble are filtered to -18dB at 50Hz, all other signals being unchanged.

Using a filter to remove very low frequencies caused by, for example, a warped record has an unfortunate tendency to remove some of the music as well. This circuit avoids the problem.

Since rumble is caused by vertical movement of the stylus, the stereo signals are anti-phase and can be removed by passing all out-of-phase components through a second-order high-pass filter to produce a bass mono signal, common-mode signals at all frequencies being left unchanged.

A modification of the biquad filter is used, in which the integrators and inverter need not be of wide bandwidth, an ordinary standard op-amp being satisfactory. Fewer capacitors, a steeper roll-off of the out-of-phase signals and more attenuation are obtained than is the case with earlier designs[1,2].

It is possible that the design may find application in the protection of the cutter head in disc mastering[3] and preventing large speaker-cone excursions at LF.

John Lawson

Cheltenham
Gloucestershire

References
1. Macaulay, J P, Circuit Ideas, Wireless World, September 1979.
2. Langvad, J, *Letters, Wireless World, March 1980.*
3. Eargle, J, *Sound Recording, 2nd edition, Van Nostrand (1980), ch.10.*

Accurate gated oscillator

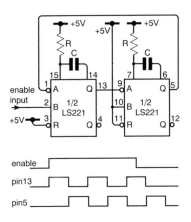

Two flip-flops ensure accurate starting of a gated oscillator.

Many gated oscillators do not start accurately; they have a tendency to produce a short or long first cycle. This circuit uses a pair of *74LS221*s to make sure the output is clean. the edge fed back from the second circuit terminating the first cycle.

With both timing circuits equal in value, frequency is approximately $1/(1.4CR)$ up to a realistic maximum of 10MHz. Differing timing circuits on each flip-flop will give other mark:space ratios.

Jeffrey Borin

Harrow
Middlesex

Fast-response PLL frequency multiplier

To avoid the effects of noise in a 50Hz-input PLL multiplier, it is common to use a long-time-constant low-pass filter, which prevents the multiplier responding to the noise. It also, however, entails a delay in the desired response. The circuit described obviates the problem.

The design allows multiplication of the 50Hz input by anything from 4 to 4000 in multiples of 4, ignoring frequency reference disturbances that last less than a second; it takes four seconds to change frequency from 200Hz to 200kHz.

When the loop is locked, the phase-comparator pulses (PCP) output at pin 1 of the *4046B* PLL is at 7.5V, going to -7.5V when it is out of lock. On changing division ratio in the programmable counter, loop response is slowed by the filter from pins 9,14

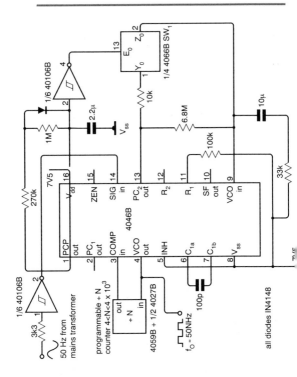

Using pulse comparator output of PLL to switch filter in and out of multiplier circuit avoids long wait for lock while avoiding spurious frequency changes due to the effects of input noise.

to V_{ss} and if the loop is not locked after 2s, switch 1 closes and shorts the filter; it then stays closed until the loop has been in lock for more than 0.5s. This delay prevents overshoot and the hysteresis of the closing and opening of the switch enables the loop to track intended frequency variations at the input without imposing a long lock time for large changes.

Richard Pulham

Use a printer port for general i/o

Printer ports on IBM PCs and look-alikes are
Centronics parallel interface standard and can be

0378H	output (data port) pins 2-9
0379H	input five lines (D3-D7) pins 15,13,12,10 and 11
037AH	output four lines (D0-D3) pins 1,14,16 and 17. D4 enables
IRQ7	
037AH	input four lines (D0-D3) bidirectional
0378H	input; pins 2-9 can only read data on data port

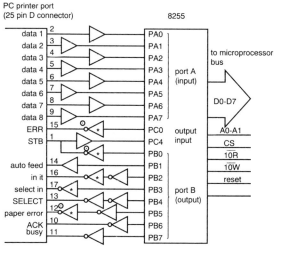

PC printer port
(25 pin D connector)   8255

1 pull down 47k at data 1 to data 8
2 * open collector inverter 7405/06
3 inverters CD 4049 and non inverting buffers
  CD 4050
4 ⊙ pull up 4k7

*Figure 1. PC printer port used for external data
reads. This 8255 PPI handles data transfer under
control of micro.*

used to read byte-wide data from an external source. Assuming a base address 0378H for the set of five ports in the standard, the table gives address and direction.

Input and output lines of 037A share the same pins. Output lines of 037A are buffered by open-collector gates and may be kept off by setting 04 on this port. In this condition, these four lines and the five at port 0379H can be used for the external data, with one line available for direction testing.

Input data is read at two ports and possibly in one 16-bit read operation. Bits are compiled to organise the input data byte; in some cases, inverters are

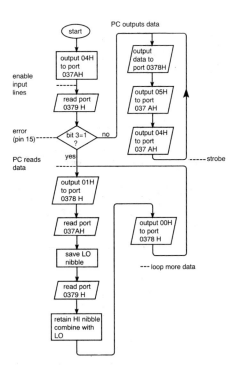

Figure 2. PC flow chart for data reads.

needed. Figure 1 shows an interface using a programmable peripheral interface *8255* controlled by a microprocessor to handle data transfer. The PC tests direction by sensing the ERR line.

Flow charts for both PC and host are at Figures 2 and 3.

R N Misra

Physical Research Laboratory
Navrangpura
Ahmedabad India

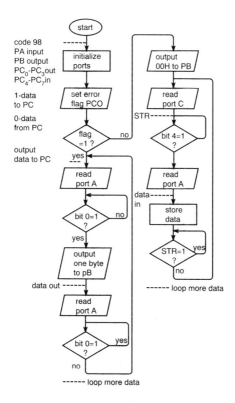

Figure 3. Host transfers data to PC.

Diode probe thermometer

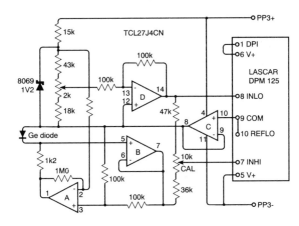

*Diode probe thermometer with digital readout
measures from -25°C to 125°C5.*

I have based this design on that by Henderson
(*Wireless World*, June 1981, p. 50) to exploit the low
drift and low current needs of the Texas TLC series
op-amps. It uses an inexpensive 3.5-digit, 200mV
display to give a stable 0.1°C resolution reading of
temperature from −25°C to 125°C.

  Op-amps A and B compose a constant-current
source for the probe, defined by the 8069 bandgap
reference and 1.2kΩ resistor, B also providing the
positive meter input by way of the calibration pot.
Op-amp *C* buffers the 0V rail to the circuit, keeping
it at 2.8V below battery voltage as determine by pin
nine of the DPM (if other DPMs are used, this
voltage may need to be obtained by other means).
Zeroing is via op-amp D.

  I used cermet multi-turn pots and 1% metal-film
resistors. If a silicon probe diode is used, the pot

chains may need modification. Calibrate the instrument at freezing and boiling points of distilled water; the zero pot adjusts the freezing point reading, to be set first, and then the calibration pot for the boiling point.

The diode cathode goes to the probe tip for quick response. A length of thin-walled 3mm OD tube (an old telescopic aerial) forms the probe shaft with a length of 36SWG wire soldered inside to pass through the tip with the diode for the cathode connection. The other end takes a 2.5mm phone jack.

H Maidment

Salisbury
Wiltshire

Power for car audio

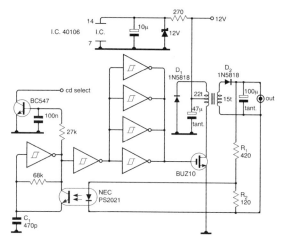

Flyback converter with opto-coupled feedback gives floating power supply for car audio equipment.

Separating signal and power supply grounds in car audio equipment can present a problem, a frequent solution to which is to provide a kind of floating supply by means of a switching PSU. The circuit shown has a true floating output which is used to supply 700mA at 5V to a portable CD player.

Voltage feedback from the flyback converter output is taken by way of an opto-coupler, the values of R_1 and R_2 being chosen to suit the NEC *PS2021* device. The *BC457* open-collector stage signals power on in the CD player to switch the audio path with no loss of efficiency, since it monitors oscillator duty cycle. Capacitor C_1 may be increased to work with other toroids at lower frequencies.

A prototype oscillates at 40-65kHz, depending on the load; regulation and stabilisation are better than 5% and efficiency is about 70%.

Paolo Palazzi

Cervignano

Italy

Simultaneous insertion and return loss plots

Using an analogue simulator to plot branch currents and node voltages of a network can provide a plot of insertion loss or gain as a function of frequency. To obtain the driving-point impedance and therefore reflection coefficient p and return loss 20log≥ ρ≥ needs more computation. It is, however, a simple matter to model a return-loss bridge at the relevant port, the bridge acting as a generator to plot insertion loss and return loss at the same time.

Figure 1 shows the model. In this form of Wheatstone bridge, voltage across the horizontal V_{32} is in proportion to the reflection coefficient relative to R_o at Z_x:

$$V_{32} = [(Z_x - R_o)/(Z_x + R_o)](V_o/8) = \rho V_o/8.$$

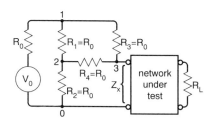

Modelling a return-loss bridge at the relevant port allows plot of insertion loss and, without further computation, return-loss plot simultaneously.

The network sees the bridge as a voltage source $V_o/2$ and R_o, corresponding to the ideal return loss bridge with 6dB insertion loss. To plot insertion loss, there must be a source of 2V instead of $V_o/2$, so $V_o = 4$ and $V_{32} = \rho/2$. Adding a voltage-gain source with a gain of 2, feeding it with V_{32} and taking it to an unconnected node gives a voltage equal to ρ; a decibel plot of this node voltage gives return loss.

C J Hall

Giubiasco Switzerland

Slow ramp generator

Ramp times of more than a few seconds have a tendency towards variability of time constant, since the capacitors used are electrolytic and therefore relatively variable.

To make a ramp with a period of 16384/clock frequency, when a five-bit resolution is acceptable, use the circuit shown here. All the resistors specified can be had in 1% tolerance to give a linearity to within 1LSB.

Since the *4020* is a 14-bit ripple counter, switching glitches are evident and are smoothed by the 10nF feedback capacitor, which may be larger if required to eliminate the steps in the output. Ramp amplitude is controlled by the 5kΩ variable resistor. If you need the ramp to drive a comparator, you can leave out the amplifier altogether and simply put a smoothing capacitor on the comparator input.

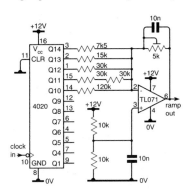

Simple circuit to give slow ramps, in cases where an electrolytic timing capacitor would be on the vague side. The 4020 is a 14-bit up-down ripple counter.

As shown, the output is a negative-going ramp, which may be inverted by using inverted *4020* outputs; and to further exploit the circuit, the counter could be replaced by an up-down counter to give sawteeth or triangles.

A H Millar
Witney Oxfordshire

Rechargeable battery tester

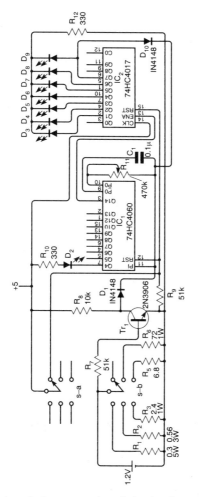

Circuit to indicate capacity of nicad rechargeable batteries in terms of percentage of normal capacity. Batteries are discharged at one-hour rate.

As a nicad rechargeable battery ages, its capacity decreases and its time to recharge increases. This circuit tests the capacity of a battery.

Fully charged, new 9V, *AAA*, *AA*, *C* and *D* batteries have capacities of 0.1, 0.18, 0.5, 2.2 and 4AH respectively; the switch selects the appropriate resistor to discharge any of these types at the one-hour rate. The *4060* is a 14-stage binary counter with a built-in oscillator whose frequency is determined by R_{11} and C_1, adjusted so that Q_{14} emits a pulse every 12 minutes to the *4017* decade counter. At the same time, Q_4 supplies a 1.4Hz signal to the led D_2, which shows that the circuit is in operation.

Oscillation is under the control of the battery voltage; if that is higher than 0.9V, Tr_1 saturates and holds D_1 off – if it is low enough to cut off Tr_1, D_1 comes on and stops the oscillator and therefore the signal to D_2.

At this point, D_{3-9} show the battery's capacity of between 20% and 140% in 20% steps. When D_9 is on, its Q_7 drive from IC_2 stops the oscillator by means of D_{10}, so that D_9 is on continuously when the battery voltage is at least 40% higher than normal.

Yongping Xia

Torrace
California
USA

50W "Blomley" amplifier

Stirred into action by the letter from W Groome in the January issue, Hans Hartsuiker presents his own design, which is based on the Peter Blomley amplifier from 1971. Mr Hartsuiker built this one in 1984 and has been pleased with its performance on

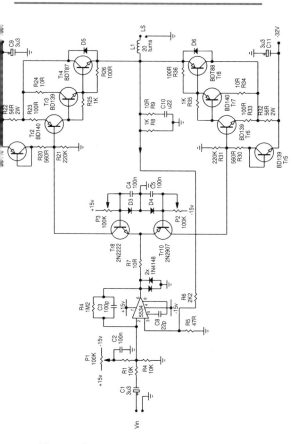

Amplifier employing the ideas put forward by Peter Blomley 20 years ago, in which phase splitting is carried out in an early stage. Output transistors are permanently conducting.

Quad *ESL-63* electrostatic speakers. It is based on the original proposition that Class-B switching should be done early, rather than in the output stage.

To take advantage of advancing op-amp design and to reduce the component count, the low-noise 5534 is the gain stage, R_4 and $C3$ determining its gain and frequency response; the two diodes protect $TR_{9,10}$ during clipping. Transistors $Tr_{9,10}$, which must be switching types, compose the phase splitter to drive the output triplets formed by $Tr_{2,3,4}$ – "a supertransistor" with the load in the collector which, together with Tr_3, takes the form of a current mirror. Inductor L_1 and the filter at the output reduce the possibility of RF into or out of the amplifier.

Points in the design's favour are that no emitter followers are at the output, with capacitive loads in mind; that all the bipolar transistors are current amplifiers; and that the output-stage devices are always on. Against, there is the onset of hard clipping; the fact that this is an inverting amplifier; and overall gain is about 48, chosen for stability; feedback factor about 200.

To set up, adjust $P_{1,3}$ to maximum; set the output voltage to zero by P_1; set output current to 60mA by P_3 (33mV across R_{22}); adjust P_2 to the point where the output current starts to increase.

As regards construction , the usual precautions should be observed and the need to mount $D_{3,4}$, $Tr_{3,4,7,8}$ on the same heat sink.

Hans Hartsuiker

Eindhoven The Netherlands

RS-232C monitor without power supply

With a little work on the RS-232C serial port found on many PCs, one can monitor received and transmitted data.

Two bidirectional switches form the circuit, each composed of two opto-couplers connected in reverse sense and two diodes to connect the data to the

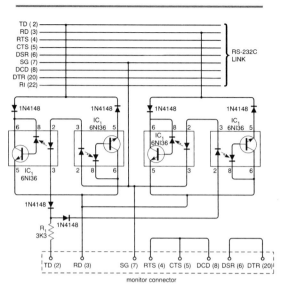

Circuit to monitor transmitted and received data from an RS-232C link without the need for a power supply. Circuit takes 2mA from monitor.

correct switches. One pair sees the TD line of the port, while the other looks at the RD line. The TD monitor is activated when pin two of the monitor connector (TD) is low, the second pair when pin two goes high to monitor RD. Since the opto-coupler leds are in series with each other and with R_1, current from pin two of the monitor connector is only about 2mA, so no power supply is needed. *6N136* couplers were chosen for low drive-current needs.

The monitor connector is an equivalent to a null modem and a microcomputer set to the same baud rate as the link will monitor the TD or RD lines. If baud rate of the monitor is at least twice that of the RS-232C link, full-duplex monitoring is possible.

Frantisek Michele

Brno Czechoslovakia

Frequency doubler

This circuit will provide frequency doubling over a wide range of frequencies.

Two monostable flip-flops contained in a *4528* use the same timing circuit – P_1, R_1 and C_1 in the circuit diagram. They trigger on opposite edges of the input waveform and, working with the Nand IC_{2a}, give an output pulse on each transition of the input. Low time constants from the *4528* allow a large frequency range.

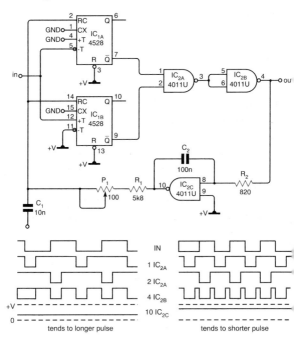

Frequency doubler working over a wide range (300-8700Hz in the case shown) and with adjustable mark:space ratio, which is constant over ±20% frequency change.

Adding the inverter IC_{2b} and integrator IC_{2c} as shown permits adjustment of the duty cycle by P_1 to give symmetry. With component values as shown, the input can vary between 300Hz and 8700Hz, and ±20% without affecting output mark:space ratio once it has been adjusted.

W Dijkstra

Waalre
Netherlands

Signals in chaos

Using a bipolar transistor as a feedback element as in Figure 1, in conjunction with a few passive components, produces a logarithmic function. Since the circuit is unstable, it is usable in a "chaotic" signal generator to give an output which, although not white noise, has a similar nature when heard.

Figure 2 shows such an arrangement. The output from the Figure 1 circuit is delayed by the phase shift of the filter and fed back to the input, whereupon a continuous oscillation is set up. Varying R_5 alters the characteristic of the basic block; some settings give a

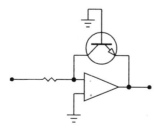

Figure 1. Circuit often described as a logarithmic amplifier, which is inherently unstable.

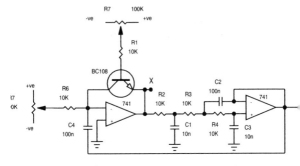

Figure 2. Chaotic signal generator uses Figure 1 circuit and phase-shifted feedback.

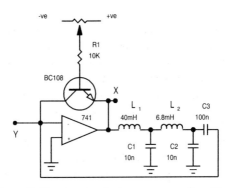

Figure 3. Simpler circuit gives same result as Figure 2, but some inductor tweaking might be needed.

periodic waveform at X, but most give a non-periodic output. Fine adjustments are possible by $R_{6,7}$, which may be omitted altogether, if desired. When set to give a "chaotic" output, the circuit is well-behaved in a chaotic sort of way and gives a continuous signal. Its behaviour becomes clear if X and Y signals are applied to the X/Y plates of an oscilloscope.

Figure 3 is a simpler circuit, but the inductor values need selection by trial and error.

D Ayers

Buxton
Derbyshire

Constant-gain tuned filter

In the classical state-variable filter, a summing amplifier is followed by an integrator train, the input, output, integral and differential of the output being summed at separately weighted inputs to give the required response. In the circuit of Figure 1, a variable element is common to both input and output terms, so that the ratio of the two is fixed at the resonant frequency at any circuit Q.

Frequency-determining components are able to take a wide range of values, since they are isolated from the rest of the circuit and do not interact with its Q value.

High-pass and low-pass characteristics are at A_1 and A_3 outputs respectively; if a non-inverting, high-input-impedance amplifier with a gain of 4 is connected to the junction of the T network, a high-Q notch is produced.

Figure 2 shows a second-order Butterworth high/low-pass filter, based on the previous circuit but with a fixed-gain summing amplifier. Both characteristics are provided and a band-pass filter of maximum gain $1/\sqrt{2}$ is obtained from A_2. To make a notch filter, feed either of the A_1 inputs to a buffer

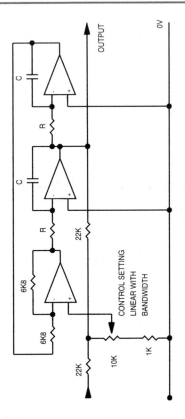

Figure 1. Constant-gain tuned filter with a wide frequency range and variable Q. High and low-pass characteristics are available and a band-pass type results from a further amplifier.

amplifier of gain $1+\sqrt{2}$. The summing amplifier gain is accurately set using preferred-value resistors; for example, 3.6kΩ and 5.1kΩ give a gain of $\sqrt{2}$ within less than 0.2%.

John D Yewen

Leighton Buzzard
Bedfordshire

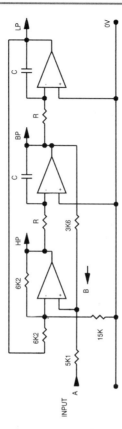

Figure 2. Variation of the Figure 1 circuit to give a second-order Butterworth high and low-pass, band-pass and notch filter outputs.

Current-to-frequency converter

As an alternative approach to the voltage-to-frequency converter circuit in which one resistor charges and discharges the capacitor, proposed by

Trofimenkov *et al* in IEEE Trans. Inst. Meas., Sept. 1986, this circuit behaves also as a current-to frequency converter in which pulse width/period ratio is a function of the input current.

When the output voltage of the *555* timer in the figure is high, C charges through the operational transconductance amplifier, in which a slight voltage imbalance at its input causes all its bias current I_B to flow in its load. This means that C is charged by the constant bias current. At the two-thirds V_{cc} point across the capacitor, the output voltage goes low and C starts to discharge through R. At one third V_{cc} on the capacitor, the output voltage again goes towards the supply and the cycle repeats. Since the input port of an OTA is a diode in series with a resistor, a voltage at the input will have the same effect as a current, so that voltage control is also possible. This circuit has been tested using a *CA3080* OTA and a *555*.

Muhammad Taher Abuelma'Atti

Dhahran
Saudi Arabia

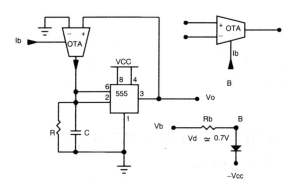

Current-voltage converter, also workable using voltage control.

Schmitt-trigger VCO

One section of a *74HC14* hex. schmitt inverter
makes a useful voltage-controlled oscillator.

When power is applied to the circuit of Figure 1,
point A is at 0V and the inverter output is high.
Capacitor C_1 charges through R_1 towards the input
control voltage V_c but, when it reaches the upper
trigger level of the Schmitt V_T+, the inverter output
goes low and discharges the capacitor through R_2 and
D_1.

As a result, the voltage at point A drops, the cycle
repeating when it reaches the lower threshold V_T. At
the output, a narrow, negative-going pulse appears,
coincident with the "flyback" at point A.

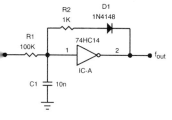

V_c (v)	f_{out} kHz
3.5	1.28
4.0	1.85
4.5	2.42
5.0	2.99
5.5	3.51
6.0	4.08
6.5	4.65
7.0	5.13
7.5	5.62
8.0	6.09
8.5	6.66
9.0	7.14
9.5	7.69
10	8.06
11	9.26
12	10.1

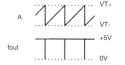

*Simple voltage-controlled oscillator using one
element of a hex. schmitt trigger inverter 74HC14.
Frequency change against control voltage is
reasonably linear, but absolute frequency will vary
fairly widely from that shown in the table, since it
depends on schmitt hysteresis, which is not held to
close tolerance.*

Pulse width is dependent on the discharge current set by the value of R_2 and the frequency on V_c, C_1 and the hysteresis of the buffer.

Yongping Xia

Torrance
California USA

Amplifier clip detector

Driving any power amplifier into clip soon proves fatal to loudspeaker drive-units. But the exact threshold of clip depends on output loading and mains voltage. Music signals are asymmetric, so bilateral detection is essential. The circuit provides true indication of clip irrespective of these variables by sensing the abrupt rise in feedback as the amplifier moves into clip.

The *SSM 2017* true differential amplifier (IC_1) reads the error voltage across the bases of the amplifier's differential input stage. R_1 and R_2 stand-off the amplifier's sensitive input nodes so stability is not affected by stray capacitance to ground. C_1 optionally prevents IC_1 responding falsely to VHF non-linearity.

In the linear mode, drive to IC_1 is close to 0V, rising sharply as the amplifier is over-driven. Unselected *2017*s have appreciable DC offset, so IC_1's output is AC coupled (C_2, R_4), then fed to window comparator IC_2, which responds symmetrically to impulses greater than ±20mV, lighting the clip led after integration (C_3, C_4). PR$_1$ adjusts IC_1's gain in the region of ×20 to ×100. It is set on test such that the led's illumination corresponds to a given THD; or a threshold between,

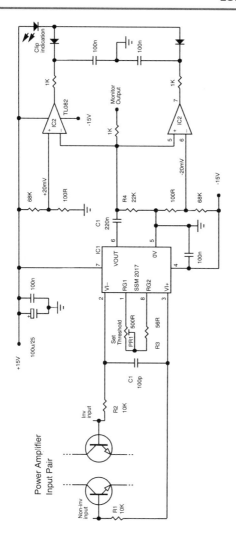

say, 0.1 and −0.6dB below clip. The monitor output feeds a scope or analyser for diagnostics.

Ben Duncan

Ben Duncan Research
Tattershall Lincoln

Programmable PWM

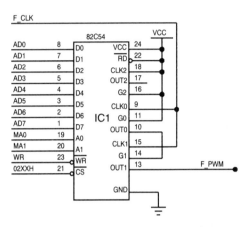

Pulse-width modulator, programmable in frequency and duty cycle, using two of the 8254's three counter/timers.

Intel's *8254* programmable interval timer contains three programmable counters, which may be made to perform the functions of programmable one-shot (mode 1) and square-wave (mode 3) generators, among others, by inputs to the D0-7 pins. In the application shown, the device is used as a programmable pulse-width modulator.

Two of the timers are used: timer 0 is programmed in mode 3 and timer 1 in mode 1, so that the output frequency is $f_{pwm} = f_{clk}/(2^n)$, in which n is the number programmed into the counter. Duty-cycle range is from 0 to (2^n-1) so that if, for example, output frequency is to be 15625Hz and duty cycle range is (2^9-1) or 511, the clock frequency is 8MHz. The listing given here will do the job.

Shwang-Shi, Bai
Chun-Shan Institute of Science and Technology Taiwan

```
timer_cntr              equ 0203h       ;timer/counter 8254
timer_0                 equ 0200h       ;timer0
timer_1                 equ 0201h       ;timer1
timer_2                 equ 0202h       ;timer2
rseg at 30h
t0_value                dsw 1           ;timer0 count value
t1_value                dsw 1           ;timer1 count value
t2_value                dsw 1           ;timer2 count value
mode_set                dsb 1           ;timer mode set
cseg at 2080h

init:
1db mode_set,#00110110b ;timer_0 as mode 3
stb mode_set,timer_cntr
1db mode_set,#01110010b ;timer_1 as mode 1
stb mode_set,timer cntr
1d t0_value,#512        ;512=2^^9 if you decide N=9
stb t0_value,timer_0    ;i.e. the range of duty cycle
stb t0_value+1,timer_0  ;of PWM is from 0 to 511
1d t1_value,#100        ;if you want duty cycle = 100
stb t1_value,timer_1
stb t1_value+1,timer_1
halt:                   br halt
end
```

Linear swept VCO

A criticism levelled at conventional audio sweep generators is that no matter how pure the sine-wave from the VCO may be under static (non swept) conditions, the picture changes when the VCO is swept.

The following scheme, which uses a combination of sweep and burst techniques, is a way of overcoming this.

In essence, a staircase ramp generated with a clock, an n-bit counter and an n-bit D-to-A[1]. The ramp is suitatbly scaled before being applied to the VCO[1,2]. The output from the VCO is gated off for one cycle while it slews at each step in the ramp.

A positive transition of the clock sets FF[1]. On completion of the current cycle from the VCO the

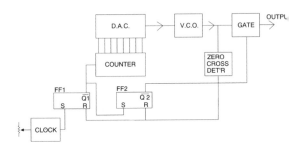

zero-crossing detector produces a positive edge which resets FF$_1$. The positive edge from FF$_1$ now clocks the counter and sets FF$_2$. Q$_2$ inhibits the output. The D-to-A output increments to the next staircase level. At the end of the next cycle (distorted as it slews) the zero-crossing detector resets FF$_2$ The ouput inhibit is removed and the circuit produces a burst which lasts until the next clock pulse where upon the above action repeats.

The ouput therefore consists of bursts of undistorted and complete sine-waves, the frequency of which increases with each step of the staircase.

DM Bridgen

Racal Radio Ltd
Reading

References
1 J, Wannamaker, *Radio-Electronics, Feb 1991, pp 43-49+70.*
2. D.M. Bridgen, *EW+WW, Feb 1992, pp 102-105.*

AC logarithmic amplifier

One fet, four diodes and a pot. form a very simple amplifier whose characteristic is logarithmic, linear or exponential, depending on the pot. setting.

Figure 1 shows the circuit with a 50kΩ adjustment, in which a central position corresponds to a linear transfer and movement to the left an exponential

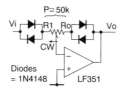

Figure 1. Simple log. amplifier with adjustment from exponential, through linear, to logarithmic characteristics.

curve. Figure 2 shows the results for some possible pot. settings, which may be varied by the addition of further resistance in series or parallel with the diodes.

C J D Catto

Cambridge

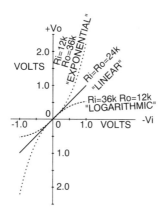

Figure 2. Transfer characteristics of the Figure 1 circuit with various pot. settings. Curves can be modified by resistance in series with or across diodes.

Increase op-amp output swing

Op-amp line output amplifiers in audio equipment are limited, in that their voltage output cannot normally exceed the ±15V rails and is usually 12-14Vpk. In the circuit shown, output is increased to 25Vpk from the ±15V supply.

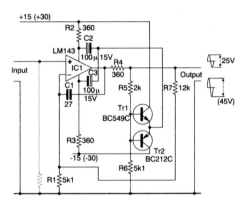

Emitter follower provides floating power supply to increase op-amp output swing by a minimum of 1.8 times, improving slew rate and leaving THD unaffected.

An op-amp, used as a non-inverting amplifier, is supplied with DC power via $R_{2,3}$ from the rails and also with a floating supply from the push-pull emitter follower. With no input to the op-amp, the supply comes solely from the rails, but with signal input, the output drives the emitter follower, which dynamically supplies additional voltage via $C_{2,3}$, increasing output swing by at least 1.8.

Input common-mode or differential input voltage limits are not exceeded, since the voltage from the emitter follower and the signal input vary in the same direction at the same time. As a bonus, slew rate is virtually doubled and harmonic distortion is unaffected.

Nick Sukhov

Kiev Ukraine

Linear current sensor

Current sensing methods that avoid the need for series resistance eliminate power loss and have no effect on the measured current itself. This circuit performs this function to a high degree of accuracy.

Figure 1 shows the system, which is effectively a phase-locked loop in which the function of the conventional voltage-controlled oscillator is assumed by the current source. The dotted line indicates a ferrite core having two windings; *W1* is part of an oscillator tuned circuit, *W2* carries the output of the current source and *W3* threads the core with the current to be measured.

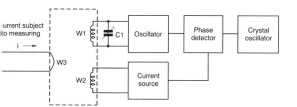

Figure 1. System diagram of current sensor that needs no series resistor. Phase-locked loop follows current in wire passing through amorphous ferromagnetic alloy.

With no *W3* current flowing, the oscillator frequency is trimmed by C_1 to equal that of the crystal oscillator, the two being compared by the phase detector. In this condition, there is no output from the current source.

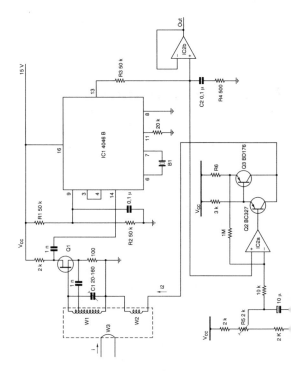

Figure 2. Circuit diagram of current sensor. ICa is current source.

A current in *W3* causes a magnetic field in the core of $H_1 = i \times W3/l_{av}$, where l_{av} is the length of an average magnetic line, causing a change in core permeability and *W1* inductance to change the oscillator frequency. The change is detected and the current source drives a current through *W2* in the opposite sense of $H_2 = i_2$ x $W2/l_{av}$. Since the loop is again balanced, $H_1 = H_2$, so that i x $W3 = i_2$ x $W2$ and $i = i_2$ x $W2/W3$. The reading does not depend on core material linearity and the measured current can be much larger than that from the current source.

Figure 2 is the circuit diagram, in which the *4046B* is the phase-locked loop, the op-amp and its output circuit forming the current source. Initially, trim C_1 to read the required level at IC_2 output, which is the circuit output and may be used for a variety of purposes such as power supply control. Potentiometer R_5 varies the demagnetising current from the current source, while R_6 sets limits to the value of this current.

Nickel-cobalt amorphous ferromagnetic alloy strip is the core material, the original being of Russian manufacture. Any material offering good temperature stability, a smooth *BH* curve and high saturation density is worth trying.

G Mirsky

Akademtekh
Moscow
Russia

Increased-resolution bar graph meter

Using only ten leds, this bar graph display resolves to 100mV from 0V to 2V. Multiples of 200mV illuminate the lower diodes fully, an increase of 100mV causing the next higher diode to flash at 3Hz, so indicating a reading above or below the half-way point in any 200mV step.

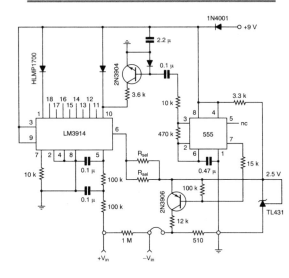

Bar graph meter reads to double its normal resolution. Next higher diode flashes, indicating above or below half step.

Reference voltage of 2V comes from the *TL431* and is applied to the top of the *3914* resistive divider; it also supplies the *2N3906*, which generates 100mV pulses when switched on by the *555*. Since these pulses are added to the floating input voltage, the comparators in the *3914* periodically go to to the next higher state if the input is midway between steps.

The differentiated *555* output also illuminates the led on pin 10 to act as a pilot light. Inhibiting the 100mV levels by means of the jumper restores the normal 200mV steps.

If both V_{in} and the jumper are removed, bias current from pin 5 of the *3914* charges the two $0.1\mu F$ capacitors to give a 20s ramp as a self test.

John A Haase

Fort Collins Colorado, USA

Stable inverter

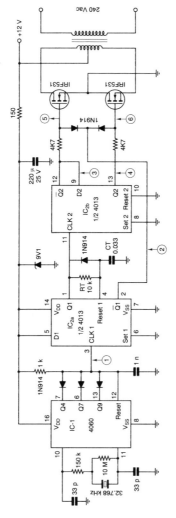

Figure 1. Simple inverter gives stable 50Hz output, accurate 50% duty cycle and 300µs dead time to protect mosfets against simultaneous conduction.

A few common logic ICs, a pair of mosfets and a transformer generate a 50Hz output with crystal stability, a precise 50% duty cycle and the necessary dead time for mosfet switching.

A 100Hz clock comes from the *4060* 32.768kHz crystal oscillator and 14-stage ripple counter, which is reset by the diode And arrangement after 328 pulses. Half a 401 D-type flip-flop, IC_{2a}, is configured as monostable by way of R_TC_T and produces the $300\mu s$ dead time (waveform 2); its output also triggers the second half of the *4013*, which is a bistable flip-flop producing 50Hz complementary gate drives for the mosfets. Rate the mosfets and transformer to suit requirements.

Connecting the And diodes to outputs Q_5 and Q_6 gives a 60Hz output.

M S Nagaraj

ISRO Satellite Centre
Bangalore
India

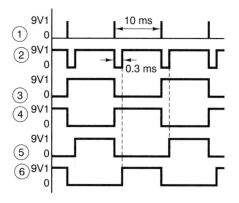

Figure 2. Timing diagram of inverter. Dead time is generated by first half of 4013 used as a monostable flip-flop.

Gyrator acts as electronic choke

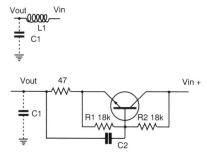

The lower circuit simulates an inducter, L1, shown here in a power filtering configuration with C1. Gyrated inductance in Henries is approximately equal to half the capacitance of C2 in microfarads. This gyrator does not require a ground reference.

A gyrator can take the place of a straight resistor or inductor in filtering out noise and ripple on a power supply bus. Like the inductor which it simulates, it passes DC current with minimal voltage drop across its two terminals while providing a high impedance to AC.

The maximum level of AC noise and ripple which it can deal with is determined by the DC voltage drop across the transistor, itself determined by the ratio of R_1 to R_2. The simulated inductance relates directly to the size of C_2 and has a value in henries of about half the value of C_2 in μF.

The gyrator may also be used to hold down a PSTN line while providing a high impedance path to the 600Ω audio.

P Strict

Reigate Surrey

High-power, class-A amplifier

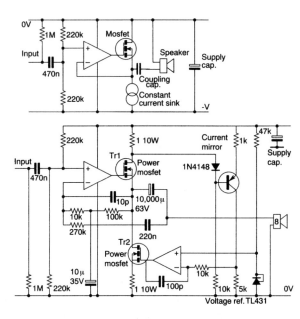

Class-A amplifier in which the driver mosfet operates in ideal mode; in which crossover distortion is negligible; and which becomes a class-AB type at higher powers.

To draw only direct current from the supply, to provide high power and to eliminate crossover distortion were the aims of this design.

Main circuitry is concentrated around IC_1, which has AC and DC feedback and provides a gain of 28 from 2Hz to 59kHz at a power output of 50W. Power transistor Tr_2 completes the DC feedback loop from the drain of Tr_1 to maintain constant current flow in the upper power transistor, taking as much current from the supply as may be needed to do that. The driver mosfet thereby acts as an ideal transistor.

A M Wilkes

Glasgow

Pulse-width detector

Pulses whose width is within an adjustable window pass to the output of this circuit.

Positive-going edges of input pulses trigger the 74121 monostable, output Q going high for a period $t_1 = 0.7R_1C_1$. This output and the input pulse go to the 2-input Xor, the monostable output also charging

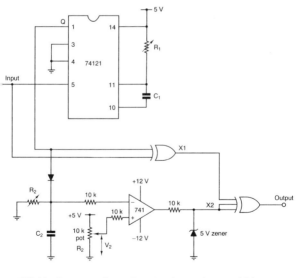

Width detector allows input pulses whose width lies inside an adjustable range to pass to the output.

C_2 to 5V. Voltage V_2 is a variable reference between zero and 5V.

While Q is high, X1 is low, X2 is low and the circuit output remains low. When the monostable times out after t_1, X1 is in the same state as the input, this being passed to the circuit output via the 3-input Xor. Capacitor C_2 starts to discharge through R_2 and, at time $t_2 = R_2 C_2 \ln(5/V_2)$ after Q collapses, X2 goes high. Since two inputs of the 3-input Xor are high, circuit output is low.

Pulses of width greater than t_1 and less than $t_1 + t_2$ pass to the output, the limits being adhustable by R_1 and R_2.

KV Madanagopal
Madras
India

Tweaking a D-to-A converter

It is possible to improve the linearity of a cheap digital-to-analogue converter by means of two resistors.

Ramping the input of the D-to-A from zero and viewing the output on an oscilloscope shows that the major non-linearity appears half-way, when the code changes from 0111...111 to 1000...000, secondary discontinuities coming at each quarter, Figure 1. Since the end of the second quarter is the half-way point, it is simplest to start by correcting first and third quarters.

Ramp up the D-to-A and select R_m in Figure 2 to make the two halves linear; then choose R_n to align the halves with each other. Resistor R_n will probably need readjustment, but the procedure is simple – if a

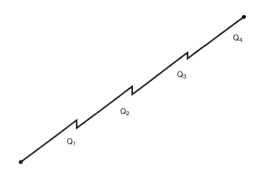

Figure 1. In general, D-to-A non-linearities appear at quarter-scale major code changes.

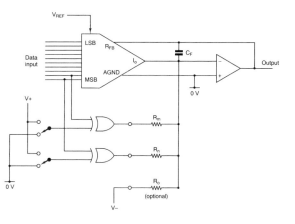

Figure 2. Simple method of linearising D-to-A converter. Resistors Rm,n compensate for largest discontinuities.

number of high-value resistors are available. Alternatively, feed static codes to the input and use a precise DVM in place of the oscilloscope; this may be more accurate, but the first method is visual and quick.

The offset resistor R_o compensates for V^+ current through the two linearising resistors and is optional. Since linearity errors can be either positive or negative $R_{n,m}$ must pull up or down; hence the Xors or possibly Nands. Do not move the non-inverting op-amp input, since this would vary gain with input-code changes.

If the D-to-A has built-in latches, the state of the two MSBs must be stored separately in a dual flip-flop updated with the D-to-A.

CJD Catto

Cambridge

Auto-reverse motor control

This one-chip circuit not only runs a motor in alternate directions for adjustable times, but stops it on reverse to avoid damage.

An oscillator based on IC_A is adjustable in period and duty cycle, R_4 controlling its output duration at high level and R_5 that at low level, both between 10s and 10min. The levels themselves control direction via IC_C, Tr_1 and RL_1. Resistors $R_{1,2}$ avoid the need for C_1 to charge from zero at switch-on, causing a longer high-level period.

Narrow pulses formed by $C_{2,3}$, $R_{7,8}$ and $D_{3,4}$ from the edges of the IC_A waveform discharge C_5 at each direction change, turning off Tr_2 and relay RL_2 to stop the motor for a time adjustable up to 10s by R_9. Resistor R_6 and C_4 delay direction changes until the motor has stopped.

Yongping Xia

Torrance California USA

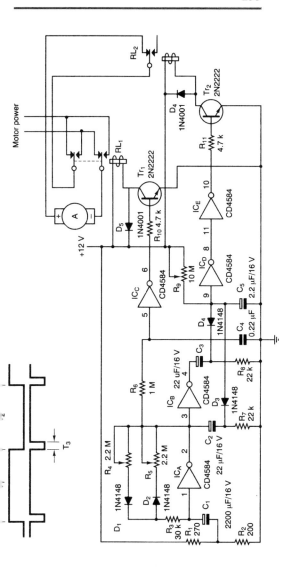

Motor controller alternates motor direction for
adjustable periods and guarantees that motor is
stopped before a direction change.

Linear ramp generator

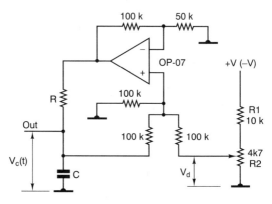

Linear, long-period ramp results from bootstrap constant-current charging circuit, adjustable by pot. setting.

Linearising an ordinary *RC* integrating circuit produces a long-period, adjustable ramp for use in timers.

As *C* charges, the voltage across it is applied to one input of the resistive summer, the other input being taken from the R_1R_2 tapping point. Voltage at the non-inverting op-amp input and, since the op-amp provides a gain of 2 to make good the loss in the summer, at the top of *R*, is therefore $V_d + V_{c(t)}$. Voltage at the lower end of *R* is $V_{c(t)}$, so that the voltage across it is $VR = V_d + V_{c(t)} - V_{c(t)} = Vd$, the bootstrap voltage providing constant current through *R* and therefore linear charging. This voltage sets the ramp time, with no *RC* timing circuit adjustment needed. A negative-going ramp is formed by a negative voltage on the voltage divider.

Valery Georg Chkalov

Oblinsk Russia

FSK receiver has auto decision-threshold control

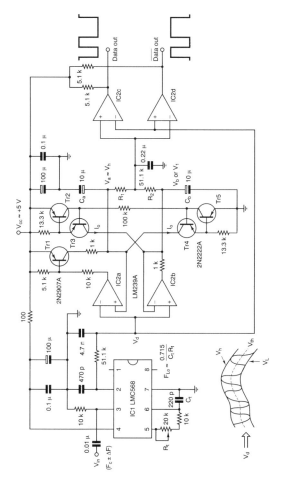

Simple method of adapting data-recovery decision-level threshold to take account of noise, drift and gain variations in the amplifiers and demodulator.

This low-cost, low-power PLL receiver for FSK binary-coded data automatically sets the best decision-threshold level for recovering the data.

Ideally, V_{th} should be midway between V_h and V_l, the upper and lower peak voltages of the demodulated signal, which are not usually stable due to the effects of drifts and gain inequalities. Signal from the *LM568* demodulator passes to comparators $IC_{2a,2b}$, which work in conjunction with current source $Tr_{2,3}$ and sink $Tr_{4,5}$ as positive and negative peak detectors, tracking variations of the peak voltages. Their outputs are combined and averaged by $R_{1,2}$ to put V_{th} half way, this voltage going to the output comparators, which also accept the input from the demodulator. Current generators improve the dynamic range of the peak detectors.

With a 100mV carrier at 110kHz, the local oscillator at $2F_c$ and frequency deviation between 500Hz and 5kHz, drift was simulated by adding a 0.1Hz signal to the 300b/s digital modulation. When $V_h - V_l$ is between 100mV and 600mV and V_{th} 1-3V, data is always recovered with negligible jitter.

G Stochino
Ericsson Fatme Spa
Rome
Italy

Cheap mosfet audio power

To provide very good performance using cheap mosfet output devices, this circuit uses an op-amp driving a conventional source follower output stage.

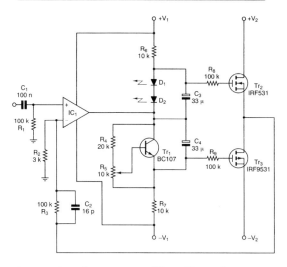

Low-cost audio power amplifier with mosfet output and an op-amp driver. IRF531/9531 cost under £1.

Transistor Tr_1 sets the quiescent current, the two leds dropping a fixed 3.2V; a single zener would be just as suitable. There is a small drift, but since quiescent current is not critical, there is no correction.

Since the output stage voltage gain is less than unity, the op-amp should be chosen to have enough internal frequency compensation to avoid instability; it should also have a high slewing rate and reasonable current drive. Capacitor C_2 reduces gain at higher frequencies.

Power output depends on supply voltage; at ±22V, power is 15W into 8Ω.

Being a low-cost circuit, it may be feasible to use several with active crossovers, rather than one higher-quality amplifier.

Andrew Southgate

Sutton, Cambridgeshire

No-loss tuned circuit

The circuit shown is (ideally) a loss-free parallel
tuned circuit, for use as part of a band-pass filter or
as a reasonably accurate sinewave oscillator if the
dotted resistor is in place, although a diode limiter
across one of the capcitors may be needed in the
oscillator form. With component values normalised
to unity, the dotted resistor would be typically 50.

Frequency control is also unity, R being the top
half to give

$$\omega = \sqrt{\frac{R}{1-R}}$$

Varying the control gives a fairly straight line over
around two octaves, accurate enough for use in a
simple spectrum analyser.

McKenny W Egerton jr
Owneys Mills, Maryland, USA

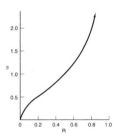

*Loss-free parallel
tuned circuit with one
frequency control to
give a nearly straight
resistance/frequency
response over two
octaves.*

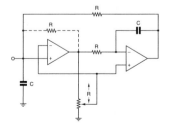

Transconductance squarer

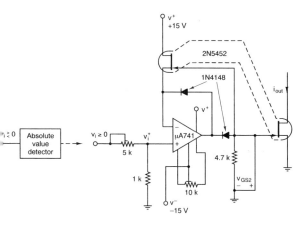

Current at the output of this simple circuit is proportional to the square of the input voltage, for inputs greater than zero. Absolute-value circuit makes it a true squarer.

An op-amp and a dual fet combine to give output kv_i^2 when $v_i > 0$.

National Semiconductor's *2N5452* n-channel dual fet has good matching between the two devices and low output conductance. Normally, the voltage between the fet gates and the non-inverting op-amp input is constant at V_p, the pinch-off voltage of the two transistors. V_{GS2} is $(v_i^+ + V_p)$ and i_{out} is proportional to v_i^2.

The coefficient k is adjustable by means of the $5k\Omega$ input variable, the two diodes and the $4.7k\Omega$ resistor ensuring that i_{out} does not exceed I_{DSS} and affording negative feedback should v_i^+ become greater than $|V_p|$. Output voltage must lie within the 5-15V range.

Including an absolute-value detector at the input produces a true squarer: either polarity of input gives the same output. With a $\mu A741$, the circuit works at several kilohertz.

Alexandru Ciubotaru

University of Texas at Arlington
Texas, USA

Electrocardiograph simulator

This simple ECG simulator has been useful for some years in teaching and equipment test and development.

$IC_{1b,1a}$ form an astable flip-flop at 20Hz to drive the clock input of the decade counter IC_2, which produces sequential highs on its Q_{0-9} outputs. The voltage at the common point of the output resistors is determined by the potentiometer action of one on resistor and nine off, the values being chosen to form the required pulse shape.

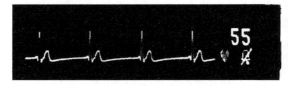

When Q_9 goes high, it inhibits the $IC_{1c,d}$ monostable for a time set by the 2.2MΩ variable resistor, so controlling the pulse rate from 30pulses/min to 160p/m. At each pulse, led D_2 blinks.

The 10kΩ resistor and 1.5μF capacitor smooth the stepped output from the resistor bank and the 10kΩ pot. sets the output to 0.2mV-2mV pk-pk. Over 4kΩ output impedance could trigger the "lost electrode" alarm on some cardiac monitors.

Alberto R Marino

Madrid Spain

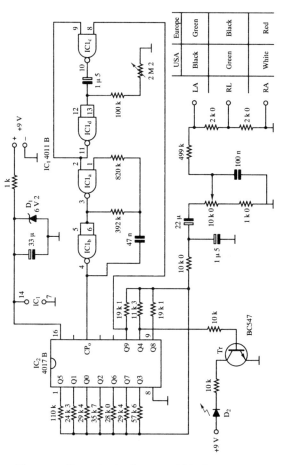

Electrocardiograph simulator, variable from 30 to 160 pulses per minute, pulse shape being synthesised by sequential outputs from decade counter and resistor bank.

Zero-crossing detector

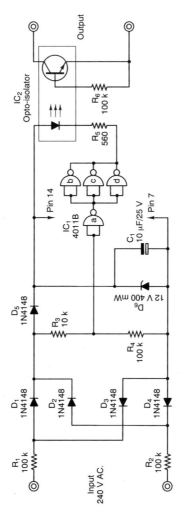

Simple zero-crossing detector produces an isolated pulse, used by the author to synchronise a sawtooth generator.

At every zero crossing, this circuit produces a 150μs pulse, centred on the crossing, the output being optically isolated from mains input.

Diodes D_{1-4} rectify the input current, which is limited by $R_{1,2}$ to around 1.3mA; power dissipation is minimal, so virtually any kind of resistor will suffice. Output from the rectifier powers the quad Nand IC_1, zener D_6 limiting the voltage to about 12V.

Since input to IC_1 is normally high, output from the three parallel gates is high, except for the period around each zero crossing, when input to IC_1 is low and so is the output, which turns the led and output transistor on. Resistor R_6 avoids trouble caused by stray coupling at the transistor base. Capacitor C_1 takes about 100ms to charge at switch-on.

AJ Flind

Taunton, Somerset

Low-loss lamp dimmer

Once having shut the car door in a dark car-park and thereby extinguished the internal light, you can't find the keyhole to lock it. But this circuit dims the light slowly until the door is locked.

Transistor Tr_2 drives the lamp and in turn is driven by the *3524* regulating pulse-width modulator. With the door switch closed (door open), C_3 is fully discharged, Tr_1 fully conducting and C_1 shorted. Pin 2 of the PWM, the non-inverting input, is high and the lamp fully on.

Closing the door and thereby opening the switch causes C_3 to charge through $R_{5,6}$, holding Tr_1 in saturation for about a minute. When Tr_1 cuts off, C_1 charges slowly through R_7, bringing pin 2 of the PWM to 0.6V, and slowly dimming the lamp.

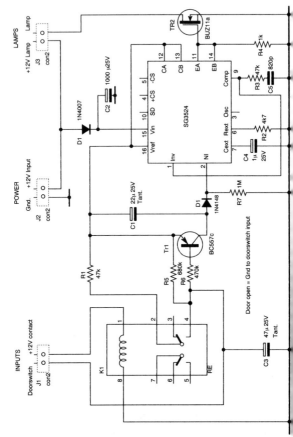

Pulse-width modulator automatically dims internal car lighting, giving time to lock the doors before the light goes out.

If matters were left there, one would have to wait for the lamp to go out completely, which would take some time, so the relay is actuated by the 12V contact on the lock to charge C_3 quickly through R_1. Diode D_1 and C_2 prevent interference when the engine is being started.

The driver transistor needs no heat sink since it runs in hard switching.

Yves Delbrassine,

Hovenierslanden 3,
8200 Bruges,
Belgium

This principle is, of course, not limited to dimming car lighting; low-loss drive for high-power loads is common, although it is not often seen described for such simple projects. Among the possibilities that spring to mind is a miniature crystal oven using the PWM to drive a power transistor – Ed.

Precision full-wave rectifier

Full-wave rectifiers usually need two op-amps; this circuit uses one.

Negative inputs force D_2 to conduct and the op-amp acts as an inverter with a gain of 0.5 (R_2/R_1). On a positive input, D_2 disconnects the op-amp from the output, which becomes the input multiplied by $R_2/(R_2+R_3)$ – again, half the input.

Disadvantages are the high output impedance (R_2) on positive inputs, and a varying input impedance. Choose the value R so that it is much higher than the source impedance and much lower than the load impedance.

A H Millar
Witney Oxfordshire

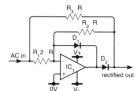

This precision full-wave rectifier uses only one op-amp instead of the usual two.

Diode protection for cmos

Low-frequency oscillators using cmos logic gates commonly need a large timing capacitor and resistor. As a result, on switch-off, an excess voltage can exist on the input terminal, although it has no effect during operation. The oscillators shown in the diagram incorporate the *CD40106* and *74HCT132*, which contain internal diodes on the input terminal to protect against overvoltage, but these diodes are only capable of passing a limited current. Permanent damage would be the result of too high a current through the diode.

To avoid such damage, diodes D_1 and D_2 are connected externally to increase current handling; low-leakage types must be used to avoid a parallel path to the timing capacitor during normal operation. On switch-off, voltage on the input pin will not exceed the safe rating for the majority of cmos circuits.

Ms Railesha

World Friends Design Group
Tamilnadu India

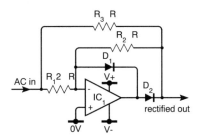

External diode augments overvoltage protection for low-frequency cmos oscillators using large capacitors.

Simple DC modulator

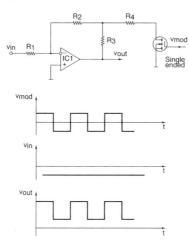

Figure 1. Modulator uses only one switching element, the signal not being isolated at any time.

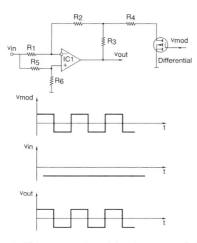

Figure 2. This connection either inverts or follows the input, depending on the modulating voltage, and gives a symmetrical output.

In addition to its requirement for only a single switch, this modulator does not isolate the input. As the feedback loop of the op-amp in Figure 1 is varied by v_{mod}, its gain changes and is given by $(R_2 + R_4 + R_2R_4/R_3)/R_1$. The output is not, however, symmetrical about zero.

In Figure 2, the op-amp forms either an inverting amplifier or a voltage follower, depending on v_{mod}, to give an output about zero.

N I Lavrantiev

Schiulkovo
Moscow Region
Russia

Smart fet battery charger

This circuit uses an external transistor wrap-around to boost the current capability of a voltage regulator for a constant voltage lead-acid battery charging application. Using the International Rectifier *IRFS3010* smart mosfet (*EW+WW* Dec 93 p990) in this position confers current, voltage and thermal protection to the circuit. Standard devices will also work but will self-destruct under short circuit and other unfavourable conditions.

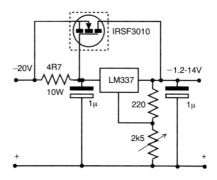

The normal configuration would use a positive variable regulator of the *317* type together with a pnp bypass. Since the IR fet is an n-channel device, it requires the use of a negative rail regulator producing a negative rail but this shouldn't be problem if the raw DC supply is made floating.

Operation is self explanatory. The cut-in point for the external fet will be determined by the value of the input resistor to the variable regulator. It should be chosen so that the voltage across it fully enhances the channel before the *337* reaches its own 1.9A current limit. The DC supply should have enough overvoltage to allow for this. Although the devices have inherent thermal protection, they require appropriate heatsinking for continuous operation.

Nick Wheeler

Sutton Surrey

1GHz frequency divider

To extend the frequency range of a 10MHz counter frequency meter to 1GHz, the input frequency must be divided by 100, to give a convenient reading. The frequency divider used in PLL tuners, shown as IC_1 in the circuit diagram, divides by 64, so that a further division of 25/16 remains to be carried out.

$IC_{2a,b}$ are a dual binary counter, which would normally count to 256, but which is reset at a count of 25 by the fed-back A, D and E outputs via And gates $IC_{3a,b}$, as shown in the timing diagram. Outputs B, C and A are further used by IC_4 and IC_{3c} to allow 16 input pulses to proceed to the output during this time, so that the division ratio is 25/16.

W Dijkstra

Waalre The Netherlands

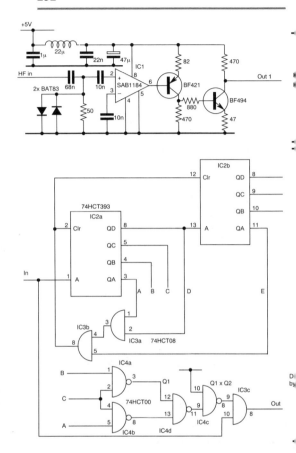

Wide-band 64 divider, followed by 25/16 divider, gives division by 100 to extend measuring range of lower-frequency counter frequency meter.

10MHz/1MHz marker generator

A common-base amplifier with a diode connected inversely across base and emitter and fed with TTL

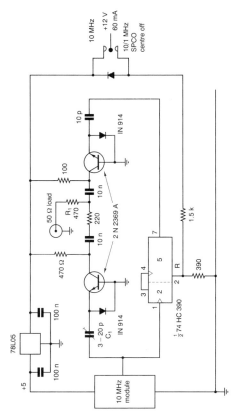

Common-base amplifiers produce fast, narrow pulse when fed with TTL edges to form a spectrum analyser marker generator at 1MHz and 10MHz.

input produces narrow negative-going pulses at the collector. Two such devices fed with 10MHz and 1MHz input generate marker pulses – in my case for a 300MHz spectrum analyser.

Signal from a 10MHz oscillator goes to the 10MHz pulse generator directly and, by way of a decade divider, to the 1MHz generator. Depending on the position of the power switch, the voltage supply goes either to the 10MHz circuit alone or to both circuits to produce a 10MHz/1MHz "comb".

Resistor R_1 isolates the load, causing a 20dB loss. The 10MHz output is −50dBm, that at 1MHz being −70dBm.

D Hutchinson

Bromsgrove
Worcestershire

Radiation detector

Microwave heating in domestic ovens and in industrial processes must comply with radiation standards. New equipment does so, at least when properly loaded, but older ovens can leak through deteriorating door and service openings, particularly when improper loading generates harmonics of the magnetron's 2.45GHz. When calibrated, this detector measures 0.01-10mW/cm^2 power density.

Figure 1 shows the simplest type of detector – a simple, half-wave dipole with a Schottky diode in the gap, of correct dimensions for 2.45GHz, although these are not critical. Depending on the diode and meter used, sensitivity is 1 to 10mW/cm^2. The slot antenna is an improvement, being as sensitive to RF power, but much less vulnerable to static charges. Adjust the distance from the oven and the angle so that a maximum can be seen. If the meter shows full-scale at 1m, it is showing up to 1W/cm^2 and that is dangerous.

Adding an amplifier, as in Figure 2, improves sensitivity to around 0.01mW/cm^2. Ceramic "radar" diodes work as well as Schottkys, whereas glass seal diodes have their problems.

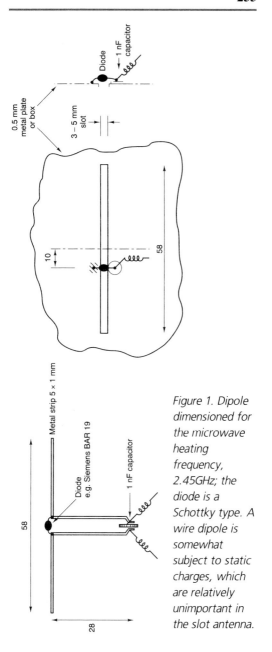

Figure 1. Dipole dimensioned for the microwave heating frequency, 2.45GHz; the diode is a Schottky type. A wire dipole is somewhat subject to static charges, which are relatively unimportant in the slot antenna.

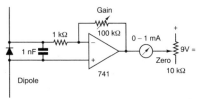

Figure 2. Adding a meter (a cassette recorder level meter was used in the prototype) increases sensitivity to 0.01mW/cm2.

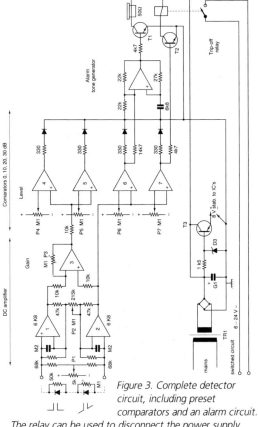

Figure 3. Complete detector circuit, including preset comparators and an alarm circuit.

The relay can be used to disconnect the power supply from the magnetron if radiation leakage becomes too high.

A more comprehensive circuit, shown as Figure 3, is provided with two detectors, one horizontal and the other vertical, each with its own amplifier. Comparators, set to produce an output when preset levels are exceeded, activate the audio oscillator alarm and trip the relay supplying power to the magnetron.

Ji í Polívka

Mexican National Autonomous University Mexico

Single pot tunes Wien oscillator

Varying one of the resistors in a Wien bridge alters the frequency and also the attenuation. In the circuit shown here, the tuning resistor also varies the gain of a compensating amplifier to compensate exactly for the varying attenuation.

Since the inverting input of IC_2 is a virtual earth, the attenuation of the bridge is determined by the setting of R_6, which is also the input arm of the compensating amplifier feedback network, the gain of which is now R_5/R_6. Since the frequency varies as $1/(R_3,R_6)$, R_6 must have a resistance range of 100:1, increasing the amplifier gain in the same proportion.

Diodes $D_{1,2}$ and R_1 form the AGC circuit. As the amplitude of output increases towards the distortion region, The diodes begin to conduct on peaks, bringing R_1 in to circuit in parallel with R_2 which, with R_4, sets the gain of the maintaining amplifier IC_1. Gain thereby reduces and amplitude stabilises. This function is usually performed by a thermistor or small light bulb with, perhaps, a little less distortion but with a certain amount of "bounce".

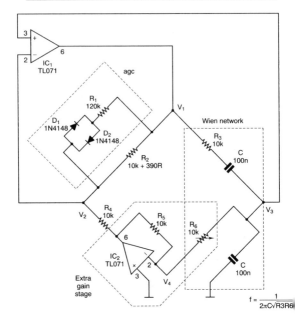

Wien-bridge oscillator with one variable resistor, which copes with varying bridge attenuation by adjusting compensating amplifier gain.

An upper limit to the frequency is fixed by the gain of IC_2 falling at higher frequencies – when more gain is needed for compensation.

W A Cambridge

Richmond
Surrey

Inductively isolated data link

Inductive coupling between two small chokes up to 6mm apart has the advantage over optical coupling

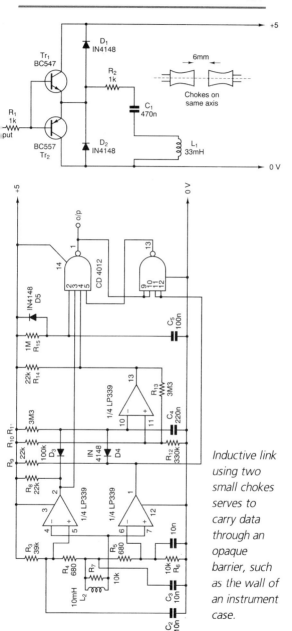

Inductive link using two small chokes serves to carry data through an opaque barrier, such as the wall of an instrument case.

that the link can be made across an opaque barrier, such as through the wall of a sealed plastic case. Inductance values shown here work for 1200baud transmission, but 9600baud should be possible with smaller chokes.

Complementary emitter-followers buffer the input and drive the overdamped *LCR* circuit $R_2C_1L_1$, in which short current pulses flow at transitions without causing any baseline shift in non-return-to-zero data. Inductor L_1, placed in line across the barrier, must be sensed not to invert the data.

Two comparators see the induced voltage across L_2 and produce low pulses for positive and negative data transitions, R_7 eliminating ringing. These pulses trigger and retrigger the flip-flop, made from *4012* gates, to reconstitute the data. During breaks in data, C_4 charges and forces the third comparator's output low, resetting the flip-flop to a known state, which is Mark for RS232. Resistor R_{16} and C_5 do the same at switch-on.

S J Kearley

Precise, wide-range capacitance comparator

This simple circuit shows whether a capacitor is higher or lower in value than a reference, or within a set percentage of its value.

Two cmos gates form an astable flip-flop, the unknown capacitor and the reference forming the timing components; unlike transistor astables, the circuit self-starts for capacitors in the range 100pF-470nF. Capacitors C_x and C_{ref} determine the on and off periods of the output square wave, whose duty cycle is therefore $d = C_x/(C_x+C_{ref})$. After the R_1C_1

filter, the direct voltage is dV_{cc}, which is compared in the Siemens *TCA965* window discriminator with a voltage derived from R_4, R_5 and R_6 (the window centre) and $R_{2,3}$ for the window half-width. Leds show the result of the comparison. In the prototype, capacitors from 100pF to 470nF were compared to within 1%.

José M Miguel

Barcelona Spain

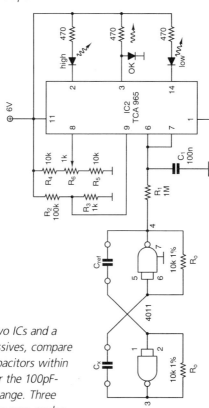

With two ICs and a few passives, compare two capacitors within 1% over the 100pF-470nF range. Three leds give over, under or within-limits indication.

Wide-range ceramic VFO

Variable-frequency oscillators using ceramic resonators and having a requirement for a specific range of frequencies are to some extent limited by the wide tolerance of the resonators. For example, at a tolerance of ±0.5%, upper and lower frequency limits on a 5MHz device vary by up to 50kHz. The circuit shown here gives a range including 4.9MHz and 5MHz consistently.

If the output is provided with a load or 10kΩ or more, or buffered, the circuit puts out 600mV from a 5V supply and stability at room temperature is ±20Hz at 4950kHz over a 10min period. A selection of ten resonators gave a 150kHz frequency coverage, including the specified frequencies. There should be no problems in using a variable-capacitance diode instead of the trimmer shown.

Paul Lovell

London

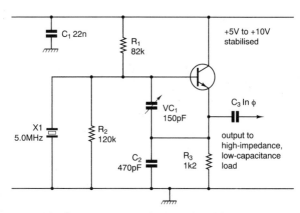

Wider frequency range than usual enables ceramic VFO to cope with wide tolerance on ceramic resonators.

Flip-flop PSU protection

This circuit will protect a series pass transistor or voltage regulator, which has access to the input of its output stage, against shorts.

Two additional transistors cross-connected in the form of a flip-flop provide the protection. If the regulator output is forced low by a short, zener D_1 stops conducting. Transistor Tr_1 cuts off and Tr_2 is hard on, disabling the regulator, which remains in that condition until, the short having been removed, the push-button is pressed and the circuit returns to normality. The led indicates the state of the circuit.

D Danyuk and G Pilko

Kiev Ukraine

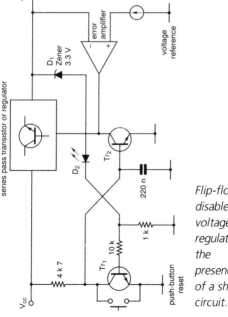

Flip-flop disables voltage regulator in the presence of a short-circuit.

Noisy video operates relay

Even when a video signal is almost submerged in noise, this circuit recognises it and operates a relay.

An input tuned circuit selects the 15.625kHz component of the signal, which is then amplified by the *BC548*, charging the 22μF capacitor. After a time determined by that process, at least 1s, the *BC213L* draws enough current to pull in the relay, the delay being necessary to prevent noise on the signal affecting the result.

There is sufficient input impedance to allow parallel connection to a video monitor without trouble. The circuit is less critical than the PLL often used for this purpose.

John Cronk (GW3ME0)

Prestatyn
Clwyd
North Wales

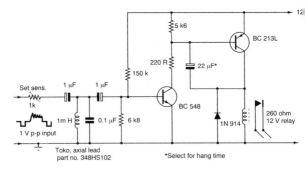

Video signals – even noisy ones – reliably operate a relay in this simple and inexpensive circuit.

Even up the marks and spaces

If your application demands a precise 1:1 mark/space ratio from an analogue drive signal of indeterminate waveform, this arrangement for steady-state signals exploits the fact that HC signals approach both supply rails.

It depends on DC feedback to the input of a couple of Schmitt-trigger inverters via an op-amp biased midway between zero and the supply voltage. Maintaining the DC average of the output signal at half the supply voltage gives the condition for a 1:1 square wave.

Ian Braithwaite

St Albans
Hertfordshire

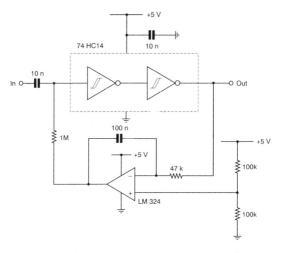

DC feedback to the Schmitt inverters keeps the output DC average at half the supply voltage, the condition for a 1:1 square wave.

12V-33V DC-to-DC converter

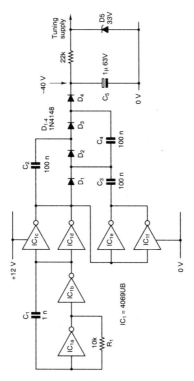

Obtaining a 33V tuning voltage for a synthesised TV tuner from the tuner's own 12V supply.

At a cost of about 50p, this circuit supplies 33V at 1.5mA to a synthesised TV tuner front end, taking the tuner's 12V as its supply.

Two buffers, $IC_{1a/b}$, make a 50kHz oscillator, which is buffered by the other four devices to provide complementary outputs driving the voltage multiplier, D_{1-4} and C_{2-4}. Zener D_5 is a safety limiter, since the tuner holds the output to the value selected for tuning.

Full-load current requirement is around 10mA. Do not short the output before the series output resistor, since this will cause latch-up and destruction unless there is a current limit in the 12V line.

Mike Harrison

White Wing Logic
South Woodford London E18

Temperature-variable voltage reference

A precise reference voltage with a controlled temperature coefficient is needed for charging sealed lead-acid cells. This circuit provides an output of 2.3V -3.9mV/°C.

Temperature-sensing device IC_1, an *LM3911* temperature controller from National Semiconductor, gives a voltage output at pin 2 of 10mV/kelvin with respect to pin 4, and incorporates a 6.8V active voltage regulator between pins 1 and 4; IC_2 is the *MAX872* 2.5V low-drift voltage reference, the resistors $R_{3,4,7}$ dividing its output to provide pin 3 of the op-amp with a very stable 2.3V. Current source Tr_1 and R_2 are controlled by the output of IC_1, these values giving a collector current of 10μA/kelvin, or 2.98mA at 298kelvin (25°C); the emitter voltage of Tr_1 is taken to the feedback input of the internal op-amp. This current in $R_{5,6}$ produces a voltage of 2.3V at pin 2 of the op-amp and therefore at the output.

Temperature changes vary Tr_1 collector current and the voltage at pin 2 of the op-amp, which varies the

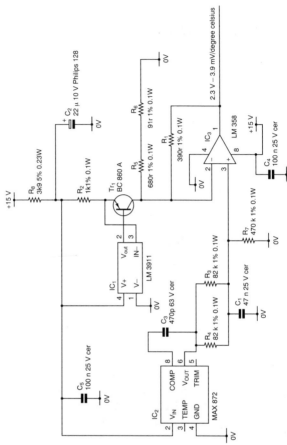

Voltage reference with a variable element due to temperature combines precision of band-gap reference at 2.3V with a controlled variation of 3.9mV/°C.

output voltage at the required rate of 2.3V less the voltage derived from a 10μA/kelvin current variation – with the values shown, 3.9mV/°C.

Kimet Rees

Brentwood Essex

RF sniffer and interference tester

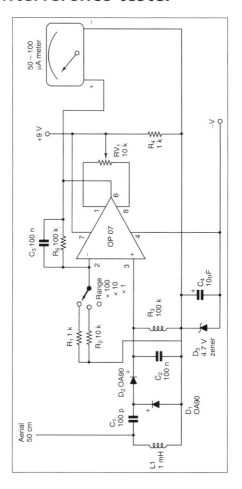

Wide-band RF detector indicates RF in the 100kHz–500MHz range at low levels. Point-contact Ge diodes confer a good frequency response and operation at low-voltage inputs.

Although intended to indicate the presence of RF emissions with the 1996 EMC Directive in mind, this circuit has been used for many other purposes, including testing car alarm keys and as a bug detector. It detects fields down to 1mW at 1m and from 100kHz to 500MHz.

In essence, it is simply a broad-band input circuit, a rectifier and meter, but for the necessary sensitivity an amplifier is needed and the diodes must be correctly chosen. Germanium diodes conduct at lower forward voltages than do the silicon type, and frequency response is higher with point-contact devices, so point-contact, germanium *OA90* diodes are the ideal choice.

A 1mH inductor on the input reduces LF sensitivity, as does the feedback capacitor. Meter zeroing is not essential, but it does allow the nulling of background signals. The meter may need series resistance to adjust sensitivity; reading is not linear and simply indicates the presence of RF and its relative level.

Alan J Jones

Newcastle
Staffordshire

Switched-gain amplifier minimises dc shift

Having its gain switched to one of two settings, this current-feedback amplifier presents excellent DC and high-frequency performance.

Bandwidth of current-feedback amplifiers is virtually independent of gain, but parasitics introduced by semiconductor gain-setting switches

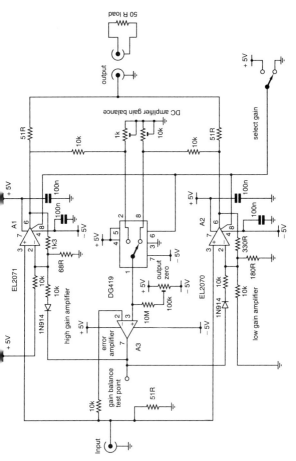

Figure 1. Switched-gain amplifier uses two current-feedback amplifiers with differing gains in parallel, the relevant one being switched into circuit using enable/disable pins.

cause the bandwidths at different gains to be unequal; miniature relays can be used, but only at low speed. Here, an *EL2071* and an *EL2070* with gains of 20 and 2.8 are in parallel, their output disable pins ensuring that only one of them is in circuit at a time. Conveniently, the disable pins are complementary, so a single gain-control line switches amplifiers.

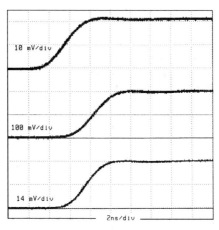

Figure 2. Step responses are virtually indistinguishable at different gains. Upper trace is an input step; middle is high-gain output into 50Ω; lower low-gain 50Ω output.

Since the DC performance of such amplifiers is poor, a number of circuit arrangements have been used to improve it. In this case, a dc amplifier is used as an error amplifier, comparing the input with part of the output. The output of the dc amplifier, which does not appear in the signal path, corrects offset in the current-feedback circuit. Correctly attenuated output signals are selected by the *DG419* switch and set by the potentiometers, the circuit arrangement being such that dc amplifier balance is unaffected by output load, which is nominally 50Ω.

The DC error amplifier corrects current amplifier offset by injecting more offset current into the inverting inputs and, since its swing must be 1V to 4V for the *EL2070* and −1V to −4V for the *EL2071*, diodes can be used to steer the correction to the relevant amplifier input. Adjust the circuit by injecting a 10kHz square wave at the input and setting the gain-balance presets for minimum AC at the error amplifier output.

Most bifet and bipolar amplifiers are suitable for the error amplifier, which determines drift; an *OP77* was used.

B Vojnovic and RA Orchard

Gray Laboratory
Mt Vernon Hospital
Northwood
Middlesex

Reference
Application note OA-7. Current-feedback Op-amp Applications Circuit Guide. Comlinear Corporation Data Book 1993-94 edition, pp 11.19-11.26.

OTA analogue divider

National Semiconductor's *LM13600* is a dual operational transconductance amplifier with linearising diodes, the bias current of which may be varied to vary the gain of the amplifier. The circuit shows a simple and accurate analogue divider using the principle. Output current is

$i_{out} = (v_{in}/10v_{in2})$mA. OTA_1 and the three transistors convert the signal voltage inputs to currents used as source and diode bias.

To adjust the circuit precisely, set R_1 with $v_{in} = 0$ so that $v_{AB} = 0$. Apply 10V to v_{in2} and adjust R_2 to give

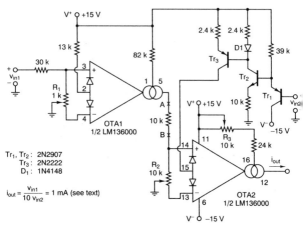

In addition to the amplifier bias input, varying diode bias on National's LM13600 transconductance amplifiers varies the gain. This circuit uses this input to make an analogue divider, giving an output current proportional to Vin1/Vin2.

zero i_{out}. Then, with equal inputs to v_{in1} and v_{in2}, set R_3 to give $i_{out} = 0.1mA$.

Signal voltage at v_{in2} must be greater than zero and $|v_{in1}|$ less than $10v_{in2}$. Ideally, v_{in1} should lie between $-10V$ and $10V$, and v_{in2} should be $1V$–$10V$. The circuit works over the 0-100kHz range.

Alexandru Ciubotaru

University of Texas at Arlington
Arlington USA

Pulse-width sequencer

Pulse trains with sequentially varied widths, and digitally programmed pulses are both generated by this circuit or a variant.

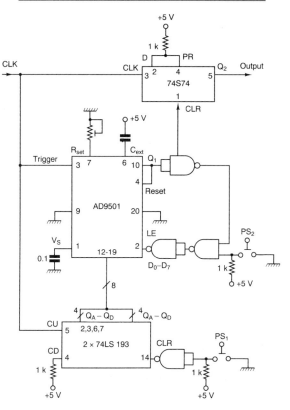

Input clock pulses produce output pulses whose widths vary sequentially up or down or whose width can be set manually.

Based on a digital delay generator *AD9501* (or the faster *AD9500*), the circuit sequentially increments or decrements the width of pulses, maximum width being set by R_{set} and C_{ext}. Input clock pulses set the flip-flop, increment the two *74LS193* up-down 4-stage binary counters and trigger the delay generator, which delays the output pulse by a period determined by the output from the counters; initially, delay is simply the propagation delay of the DGG. Also, the counter outputs are latched into the DGG to provide the new delay for the next cycle. As pulses appear at the clock input at a fixed frequency, the output pulse widths vary from minimum to maximum, whereupon pulse width returns to minimum as the counter returns to zero. Alternatively, the *193*s count down and the pulse widths decrease instead of increasing.

If dip switches replace the counters, or the counters are inhibited at the required count, the circuit becomes a digitally programmed pulse-width generator.

SR Kaul, IK Kaul and R Koul,
Bhabha Atomic Research Centre
Bombay
India

Cheaper, low-voltage ultrasonic microphones

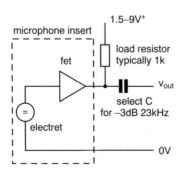

Consider using an electret microphone insert to replace the more expensive capacitor type, which needs a higher voltage supply, for ultrasonic work.

Small electret microphone inserts are sensitive to at least 90kHz, are inexpensive and only need a 1.5V to 9V supply; they can often replace the more expensive capacitor microphones which need a polarising voltage.

In the diagram, the supply comes via a resistor of 1kΩ, typically. The coupling capacitor is chosen to give a low-end roll-off in the lower ultrasonic region to avoid amplifier overload in the audio band and to provide a rising characteristic to counteract the drop in sensitivity at higher frequencies. Any further frequency-response shaping can come after the low-noise first-stage amplifier.

Les May

Rochdale
Lancashire

Automatic cable and connector tester

Connected to the parallel printer port of a PC, this device tests for shorts and open circuits in connectors and cable assemblies with up to 16 ways.

Two *4067* multiplexers take a sequence of addresses from the PC port, the top half of the 8-bit bus addressing IC_1, which switches a 5V test voltage to each of the 16 output lines in turn. While each of these lines is selected, the lower half of the bus addresses IC_2, which scans the 16 inputs for a voltage. Absence of voltage indicates open circuit, while voltage on more than one line shows short circuit.

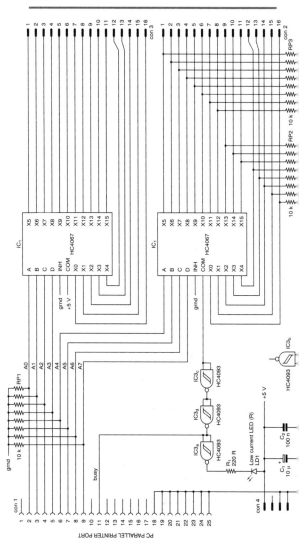

Tester connected to a PC parallel port automatically indicates open-circuits, shorts and wrong connections in connectors and cable assemblies.

When the 5V is found on an input, the printer port Busy line goes high and the led is lit. The top 4-bit nibble of the 8-bit address bus is used in the software to generate a column address, the lower nibble providing a row address to a 16 by 16 matrix, in which the BUSY bit acts as data; any lines out of sequence are thereby shown.

A G Birkett

London SE22

SCR inverter

With one or two drawbacks, the saving grace of this single-SCR inverter is its simplicity. It produces a vaguely sinusoidal waveform of around 320 AC at 400Hz, the frequency being only slightly affected by supply voltage and load. Losses in the RS 209-847 transformer dictate an efficiency of 50% with a resistive load.

At switch on, oscillations must be allowed to build up slowly on a light load, since a heavy load would cause latch-up in the conducting state, necessitating some kind of currrent limiting.

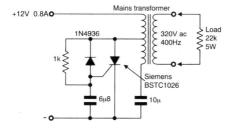

Very simple inverter, whose output is virtually unaffected by supply voltage, produces about 320V AC at 15mA.

The only variable is the resistor, which will affect the output voltage to some extent and allow lower or higher supply-voltage working. A lower frequency requires a larger resonating capacitor, which should not be an electrolytic type.

D Di Mario

Milan
Italy

OTA oscillator

In non-linear mode, an operational transconductance amplifier will function as an LC oscillator in the circuit originally described by Baxandall in 1959.

The parallel LC circuit on the non-inverting input of the 3080 OTA receives feedback from the output. At resonance, the tuned circuit appears as a resistive load $R_d = 2\pi f_0 LQ$. Bias current in the OTA, set by R_{abc}, determines g_m, which is about 2mA/V for an I_{abc} of 1mA. Oscillation at f_0 takes place when $g_m R_d > 1$, increasing in amplitude until the amplifier limits, acting as a switch to drive constant I_{abc} and $-I_{abc}$ into the tuned circuit, the waveform across it being

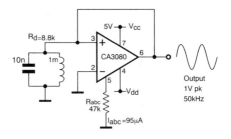

Simple oscillator for up to 100kHz, using an operational transconductance amplifier to give a 1V pk output at 50kHz.

sinusoidal with an amplitude of $4R_d I_{abc}/\pi$ and third-harmonic distortion of $(100/8_Q)\%$.

The circuit will work up to about 100kHz, although distortion increases at the higher frequencies.

J Willis

Macclesfield Cheshire

Voltage-independent time delay

Although this is a CR time-delay, no adjustment is needed and supply variations have no effect, since such variations affect both inputs to a voltage comparator equally.

Input pulses cause a "1" at the output of IC_{1b}, this being applied to the CR and to $R_{2,3}$, so that

$$V_{C1} = V_1\left(1 - e^{-t_0/R_1C_1}\right)$$
$$V_{R3} = V_1R_3\left(R_2 - R_3\right)$$

Since the delay is determined by the time needed for v_{C1} to become equal to v_{R3}, solving that equation for t_0 produces

$$t_0 = -R_1C_1 \ln\left(1 - R_3/\left(R_2 + R_3\right)\right)$$

where t_0 depends only on the CR, and voltage plays no part.

The diodes avoid the possibility of two zeros on the comparator inputs, when the input is zero. Inverting input is always high, since either IC_{1a} output or IC_{1b} output is always present.

N I Lavrentiev

Kaliningrad
Moscow Region
Russia

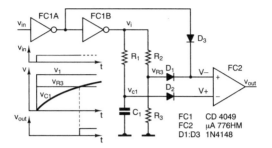

This time delay is independent of supply-voltage variations, even though the delay is determined by a CR circuit.

Comparator extends alarm system

Several alarm sensors, presenting either normally open or normally closed contacts, are used as inputs to a digital comparator, which emits an enable to an alarm system when any sensor contact changes state.

A *74LS688* has two sets of eight inputs, P and Q, and one P = Q output; when all P inputs are equal to all Q inputs, P = Q is low, otherwise it remains high.

Any Q input may be set at 0 or 1 by the dil switches and the 10kΩ sil resistors. On the P side, normally closed contacts such as that on P0 pull the inputs high, an opened switch taking the input low via its 1kΩ resistor. Normally open contacts allow transistor switches to provide normally high inputs to their P inputs, closing switches again taking the P input low; the use of transistors avoids trouble with varying supply voltages. For low-level sensors, the contact could feed the transistor base directly, as in (b).

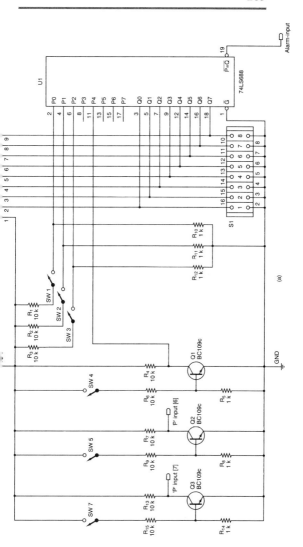

Up to eight contact sensors connected to one 74LS688 generate an alarm enable when one sensor is actuated, whether the sensors are normally open or closed contacts. At (b) is a simplified transistor feed for low-level contacts.

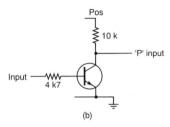

(b)

Connecting eight 688s to a summing 688 would extend the number of possible sensors to 64.

M Saunders

Leicester

Analogue switch with memory

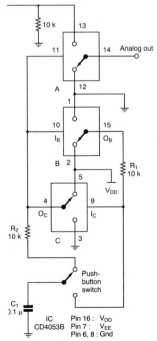

Push-button alternately opens and closes analogue switch.

A push-button switch connects or disconnects an analogue signal in successive operations, using one IC, three resistors and a capacitor.

Analogue signals go through one channel of a *4053B* three-channel multiplexer, the other two being connected as a bistable flip-flop so that outputs O_B and O_C are alternately 1 or 0. When O_C is 0, the capacitor is also at 0V; pressing the button connects it to the input I_C and triggers the flip-flop. As the button is released, the capacitor charges again to V_{DD}.

A second operation causes the capacitor to apply V_{CC} to I_C, which retriggers the flip-flop so that the capacitor is once again discharged when the button is released. Since the flip-flop is toggled each time the button is pressed, output O_C opens and closes the top channel to the analogue signal.

Requirements are V_{DD} = 5V-15V, negative supply V_{EE} = –13.5V - 0V and $(V_{DD}–V_{EE})$ = 15V maximum.

M S Nagaraj

ISRO Satellite Centre
Bangalore
India

Inductance meter

When used with a digital frequency counter, this two-transistor circuit indicates inductance within about 5% in the range $0.1\mu F$-$2000\mu H$. The reading is not direct, but a simple calculation or perhaps even a graph or look-up table gives the result.

An RF oscillator operating at 3.5-4MHz feeds a buffer to drive the counter. The terminals L_x, when shorted, result in the oscillator's natural frequency; connecting the unknown inductor increases the total

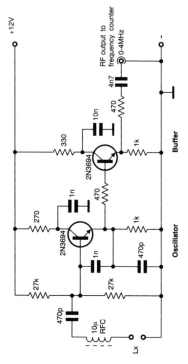

inductance and produces a lower frequency, which is displayed and used to calculate the inductance, C being known. A receiver covering 300kHz-4MHz could possibly be used instead of the counter.

The value of the RF choke must be subtracted from the calculation, which, since the total value of the two 470pF capacitors and the 0.001μF comes to 190pF, becomes

$$L_x = \left(\frac{1}{4\pi^2 f^2 \times 190 \times 10^{-6}} \right) - 10$$

where inductance is in μH, frequency in MHz and capacitance in μF.

Practically, the usual RF precautions must be observed: rigid metal enclosure; regulated supply;

good-quality capacitors; and the use of short, rigid connections, including those to the unknown inductor. Any small-signal npn transistor will suffice and the RF choke can be a standard type.

Peter Parker

Bentley
Australia

Variable-frequency generator has switchable duty cycle

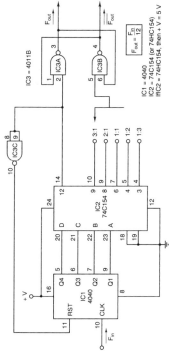

At one-twelfth of the clock frequency, this rectangular-wave generator has a duty cycle selected between 3:1 and 1:3.

At any frequency up to around 330kHz, the duty cycle of this generator is settable between 3:1 and 1:3.

Four stages of ripple counter *4040* drive the *74C154* 16-line decoder, clock input to the 4MHz counter being divided by 12. When output 12 of the decoder goes low, it sets the output flip-flop, resetting the counter, which starts to count again from zero. As the count reaches that corresponding to the decoder output selected by the switch, the flip-flop is reset. Maximum input frequency is about 4MHz.

Alberto R Marino

Madrid
Spain

Easy to use mains timer

Together with a thousand or so others, our household is currently part of an electricity tariff experiment. During the day, electricity is much more expensive than standard rate, at 12.6p a unit but in the evenings it drops to 5.6p and between 12:30 and 07:30 at night it is only 2.6p. This easy to use timer was designed to run the washing machine and dryer after 12:30 at night.

Two outlets are needed because the dryer and washer cannot be run together from one 13A socket. There is only one control, a switch, which allows manual selection of either of the two sockets. Switching to either resets the timer while switching back to the centre position initiates the timing sequence.

A red LED signals the timer's wait period while a green one signals the end of the cycle. Set up as

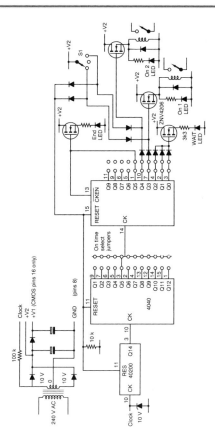

shown, the first socket turns on about 3h 40m after reset and stays on for just under an hour and a half, after which the second socket turns on for the same period. These periods mean that the timer can be initiated at any time between 8:53 and 12:55 in the evening. Using jumpers allows the on and delay periods to be altered.

Reset must be initiated in both switch positions to eliminate the possibility of having both sockets on at once, overloading the outlet. How long the capacitor can supply the CMOS ICs is difficult to determine since connecting a voltmeter to the supply increases the discharge rate. It should be a at least few minutes.

James Stevenson
Newcastle-u-Lyme
Staffordshire

Soft-start filament driver

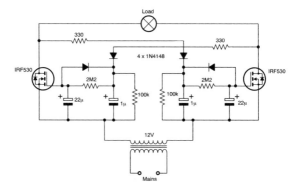

Slowly rising gate voltage on these power mosfets, acting as variable resistors, removes the current surge into a filament and prolongs its life.

Two power mosfets with a ramped gate voltage switch on slowly and eliminate a current surge into a cold filament.

The mosfets act as variable resistors, two charge pumps applying a slow-rising DC bias to the gates. Initially, the mosfets are off and the 22μF and 1μF capacitors charge through the 330Ω resistors and the internal drain/source diodes of the opposite mosfets.

During this time, the load receives drive from one mosfet and the substrate diode of the other during each half cycle. When the capacitors are fully charged, both mosfets are on, voltage drop through the circuit being determined by the mosfet $R_{DS(on)}$ of about 0.2Ω. The gate drivers give a slow charge, fast discharge characteristic.

Mosfets without the substrate diode can be used, but must have a diode such as the *1N4002* between source and drain.

Joel Setton

Crolles
France

Capacitance ratio meter

With only two ICs and without the need for a regulated supply or a crystal clock, this arrangement measures the capacitance ratio between two components to within 0.3%, presenting the reading digitally.

A latch-up-free astable flip-flop, made up of the four gates G_{1-4}, produces complememtary square waves, the on times of which are independently determined by C_x and C_{ref}. At switch-on, the two capacitors are at zero and QA and QA\ are both 1, the second flip-flop accepting this input condition and toggling one way or the other indeterminately. Assuming that QB is 0 and QB\ is 1, QB\ charges C_{ref} to the supply voltage by way of D_1. G_1 and G_2 inputs are now 1 nd 0 respectively and the outputs become QA and QB\ = 0, QA\ and QB = 1. Cx now charges and Cref discharges via R_1, this voltage toggling the flip-flop when it reaches the threshold of Set A. The new state discharges C_x and the cycle repeats.

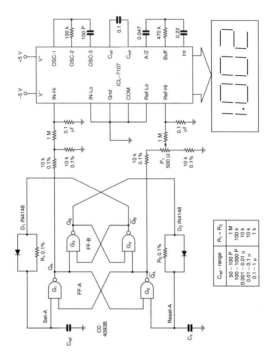

Capacitance ratio meter with a maximum reading of 2:1 and a range of 50pF-1μF. Accuracy is 0.3% and no voltage regulation or clock input is needed.

The *ICL7107* 3.5-digit A-to-D converter now indicates the ratio of integrated average voltages at the high and low inputs, its "thousands" decimal point being turned on. A potentiometer on the reference input is set so that the display shows a reciprocal reading when the capcitors are interchanged.

M S Nagaraj

ISRO Satellite Centre
Bangalore
India

Binary-to-BCD converter

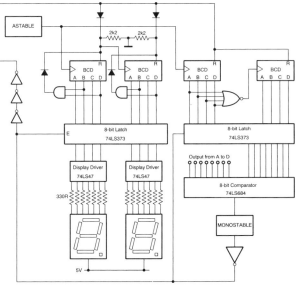

Clocking BCD and binary counters together and displaying the BCD output effectively converts binary to decimal, allowing the display of, for example, the output of an A-to-D converter. As it stands, the circuit is suitable for a 99-count input.

Converting an eight-bit binary input to binary-coded decimal can, on occasion, be useful in showing the output of an analogue-to-digital converter in decimal form.

In essence, the circuit clocks binary and BCD counters together until the output from the binary counter equals the input from the binary A-to-D converter, only the BCD output being indicated.

A 555-based clock at 7.3kHz drives both sets of counters to a maximum of 99 counts or until equality between the inputs from the A-to-D converter and the binary counter is detected by the *684* comparator. At this point, the comparator triggers a monostable flip-flop, which latches the BCD counter output so that the count is displayed. After a short delay, the same signal resets the counters, which remain reset and the count displayed for about 1s until the monostable pulse finishes, when the process recycles. Very little flicker is evident, since the process is fast.

To avoid the circuit cycling when the count is correct, a second latch between the binary counter and the comparator, so that a constant equality signal from the comparator and monostable ensure that the display only updates when the input from the A-to-D converter changes.

Richard Maggs

Rhiwbina
Cardiff

Special-purpose 74-series binary-to-BCD parts are a simpler solution for two digits but this circuit may prove more convenient as the number of digits increases – Ed.

Car radio loop aerial

This windscreen loop aerial is suitable for long and medium waves and VHF FM. An existing whip can be left in situ or an extra wire can be run parallel to one leg of the loop.

The 100nF capacitor, series inductance of L_2 and the loop tune the aerial to long wave, while medium waves are tuned by L_1 and the capacitance of the coaxial cable from the aerial to the radio. For VHF,

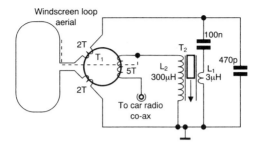

Windscreen loop aerial foils vandals and improves performance.

the ring core T_1 has an outside diameter of 6.5mm and T_2 is the oscillator coil from a MW receiver, with the slug fully in.
There is a performance improvement over a whip, particularly in tunnels.

D Di Mario

Milan Italy

High-voltage, current limited power supply

This inexpensive supply was designed to test the breakdown voltage of components and high-voltage semiconductors, using switching techniques to supply current-limited outputs from 100V to 500V at low dissipation. The output voltage is unregulated, ramping up until the set current limit is reached.

Input is up to 500V of raw DC, any reservoir capacitor being removed and possibly reused as C_{res} at the output. A switch at the input would allow the test voltage to be applied after connecting the device

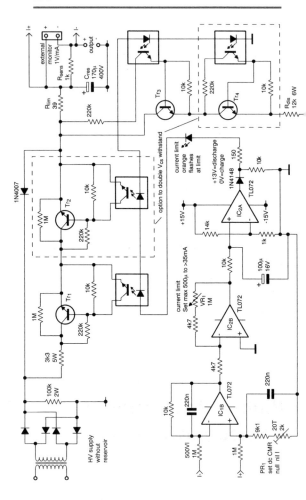

under test. Collector/emitter voltages of $Tr_{1,3}$, or $Tr_{1+2, 3+4}$ if the the extra devices to double the voltage rating are used, must withstand the input voltage but, since all the transistors are either on or off, dissipation is low and safe operating-area derating is unnecessary; TO-92, E-line or TO-126 packaged devices may be used. Collector-current ratings should be well above the current limit. The reservoir capacitor C_{res} should have low equivalent series resistance, R_{lim} being set to limit inrush current.

Amplifier IC_{1b} reads the voltage drop caused by output current in R_{sens}, which can be varied in value to allow current limits other than 20mA. Although common-mode voltage is irrelevant in a floating supply, the input to the amplifier is attenuated to allow grounded HT supplies, IC_{1b}'s ±10V CM limit coping with ±500V at the input if the 1MΩ resistors are properly rated.

Gain of the second amplifier sets the current limit and its 3MHz gain bandwidth product limits servo loop bandwidth and rate to around 15kHz at a gain of 200. Used as a comparator, the third amplifier controls the opto-couplers.

Ben Duncan

Lincoln

High-voltage current limiter/stabiliser

The basic circuit shown in Fig.1 is a current limiter using a junction fet, which operates according to

$$I_{lim} = -V_{gs}/R > I_{ss},$$

in which $0 < -V_{gs} < -V_{gs(off)}$.

For higher power, the circuit of Fig. 2 is better, the mosfet being fitted with a heat sink. The mosfet holds off the supply voltage and R_4 is a current sensor, the voltage across it driving the bipolar transistor base to apply feedback round the circuit. A 10V zener provides a supply to $R_2 Tr_1$ and, since it is connected to the top of R_4, takes into account the current in R_1.

Tr₁ IRF612 on 2°/W heatsink
Tr₂ 2N3704
Z₁   BZY88C10
Z₂   TL430C

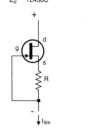

Figure 1. Essentials of a fet current limiter, where R4 senses the current.

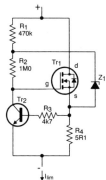

Figure 2. A mosfet handles higher power, the bipolar device providing feedback to the mosfet base.

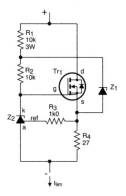

Figure 3. For improved current stability and higher output impedance, the programmable zener replaces the transistor. From the table shown, an estimate of output impedance is 1MΩ. Note the 3W resistor at R4.

	voltage	10	15	20	35	50	100	110
Figure 2	current	94.0	96.5	97.5	98.5	99.0	100.0	100.1
Figure.3	current	67.0	93.0	99.5	100.0	100.05	100.1	100.1

As the circuit stands, the current limits at 100mA, with an output impedance of around 100kΩ and with a maximum applied voltage of about 175V although, since 1000V mosfets are available, the circuit can be scaled up.

Replacing the bipolar transistor with a programmable zener gives better stability and a higher output impedance of about 1MΩ, both the quoted impedances being estimated from the performance table shown.

CJD Catto

Cambridge

A simple way to high-pass and band-pass filters

Using only a low-pass filter consisting of an *R* and a *C*, this circuit will function as an active high-pass filter and, with the addition of a similar circuit with other values, a band-pass filter.

Inputs to the *AD620* instrumentation amplifier are direct *and* via the *RC* low-pass section. At low frequencies, both inputs see the same signal, since the *RC* filter passes the signal and there is a common-mode input to the amplifier. Higher frequencies are attenuated by the filter and the amplifier receives the difference, which is amplified.

Effectively, the output of the circuit is the amplified output of the derivative of the *RC* section:

$$\frac{V_2}{V_1} = A\left(1 - \frac{1}{1 + s\tau}\right)$$

$$= A\frac{s\tau}{1 + s\tau}$$

where $\tau = R_1 C_1$ and s is the complex frequency variable.

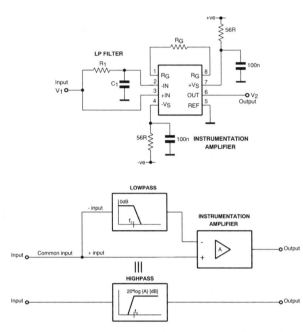

One RC low-pass section followed by a differential amplifier becomes a high-pass filter and can be made into a band-pass type.

A band-pass filter can be made by inserting another *RC* section, preferably in the common input before the original filter to give the usual low-pass/high-pass cascade, or in the non-inverting lead. These arrangements give different results and, if the cut-off frequencies are identical, the second connection gives zero output, while the first still behaves as a band-pass filter.

The circuit has found application in the amplification of a small, higher frequency signal superimposed on a slowly changing one.

Jaroslav Chum

Geophysical Institute Prague, Czech Republic

Digital sinewave generator

As an alternative to using a look-up table, an analogue-to-digital converter and a fixed filter, this circuit possesses the advantage that the frequency of the programmable output filter varies automatically.

The clock input drives the *MAX29x* filter and, via a divide-by-10 counter, a second divide by 10 counter that develops a voltage across R proportional to the count. Output frequency is $f_{in}/100$.

Lee Szymanski

Stamford Lincolnshire

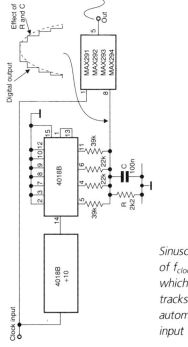

Sinusoid generator of $f_{clock}/100$ in which output filter tracks automatically with input frequency.

Spare inverter converts 5V to ±15V

If you have only a 5V rail and need a dual 15V supply, use this circuit to generate it cheaply.

A spare Schmitt inverter from, say, a *7414* operates as a free-running multivibrator at a frequency of about 100kHz using a resistor and capacitor with the values shown. As the transistor is driven on and off by the square wave from the oscillator, spikes of about four times the supply voltage develop across the 1mH primary of the 1:1 pulse transformer. Diode D_2 rectifies the spikes, which are filtered and regulated to give +15V, the current supplied being determined by the capabilities of the 5V supply and the wire gauge of the transformer. Diode D_1 rectifies the transformer output to form a −15V rail.

V Lahkshminarayanan

Centre for Development of Telematics
Bangalore I
ndia

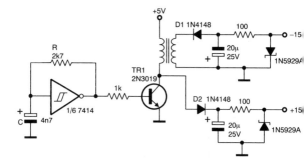

Instead of using a DC-to-DC converter or a 555 oscillator, use a spare Schmitt gate as an oscillator in this 5V to ±15V converter.

Inrush current limiter

Most of the methods of dealing with the inrush
current into large smoothing capacitors have their
disadvantages, whether they are to do with
inconvenience, performance, reliability, size or cost.
This circuit uses the high-impedance control and
large switching safe-operating area of mosfets to do
the job, with none of the above drawbacks.

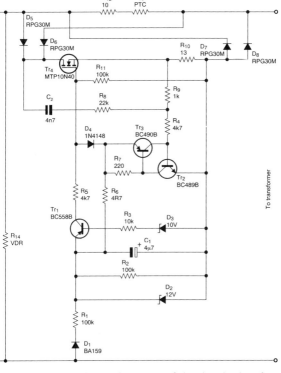

*Inrush current limiter has none of the drawbacks of
conventional solutions, such as relays, thermistors
or thyristors, and can be made part of the diode
bridge in a switching supply.*

It does not suffer from dv/dt limitations and works from zero current up to designed limits, it needs a small hold current and is proof against shorts if R_{12} is a positive temperature-coefficient type or has one in series. Under test without a PTC resistor, Tr_4 survived a short but R_{12} blew.

Resistor R_{12} determines the inrush current and R_{10} sets the maximum peak current allowed in steady-state conditions — about 4.5A with these values. A smaller maximum current would allow the use of a smaller mosfet, but would call for a higher-value resistor and more heat, so a small R_{10} is preferable.

In a switching supply, the four *RGP30M* diodes could be part of the diode bridge, the whole circuit preceding the smoothing capacitors.

Kristen Ellegård

Oslo Norway

Square waves from a 555

In the usual *555* astable oscillator, the timing capacitor charges through two resistors and discharges through one of them, the two time constants therefore being unequal. In this circuit the charge/discharge paths are similar, giving a 50:50 mark:space ratio.

The capacitor C_1 charges and discharges via VR_1 and R_2 and it is necessary to ensure that D_1 and Tr_1 base/emitter diodes are similar to avoid timing errors. Charging takes place when pin 7 is high, turning Tr_1 on; when it is low, the capacitor discharges through the diode. To obtain accurate square waves, the 5-turn potentiometer VR_2 varies the comparator control voltage. Either mos or bipolar 555s work in the circuit shown, but the bipolar

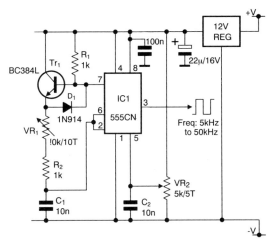

Circuit around Tr1 ensures that C1 charges and discharges at the same rate, giving accurate square waves up to 2MHz.

version gives a lower maximum frequency; the *555CN* works up to 2MHz.

To set the waveshape at very low frequencies, temporarily replace a large C_1 with a small one, set VR_2 to obtain unity M/S ratio and replace the larger C_1.

I C Rohsler

Harborne Birmingham

High-torque position servo

Parallel-connected power mosfets in an H bridge, driven by an *SG3731N* pulse-width modulator, form a simple, high-torque servo driver for a 12V, 380W DC motor.

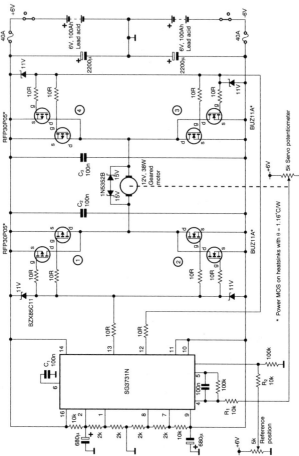

PWM IC controls mosfet H-bridge in a high-torque position servo driving a 12V, 380W DC geared motor.

Pairs of *BUZ11A* and *RFP30P05* complementary mosfets are common drain connected to simplify gate driving and in parallel to obtain the necessary current. All the circuitry is supplied by two 6V, 100Ah lead-acid batteries.

As the motor turns, it drives the 5kΩ servo potentiometer, from which a voltage is taken to one input of the PWM, where it is compared with the reference input. For clockwise rotation, the *SG3731N* maintains mosfets 3 in conduction, while switching mosfets 1 and 2 on and off. For the other direction, mosfets 2 are on and mosfets 3 and 4 go on and off. The gain of the PWM's difference amplifier can be altered by selecting new values for $R_{1,2}$ to suit different geared motors.

Capacitors $C_{2,3}$ reduce the effects of lead inductances and should be kept close to the mosfets, as should the back-to-back zeners across the motor, which absorb high-voltage spikes.

M T Iqbal

Rutherford Appleton Laboratories
Didcot

Monitor shows three-phase sequence

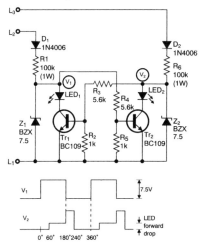

Two of the phases in a three-phase supply have a 60° phase difference with respect to the third, but in an unknown order. The monitor shown indicates this phase sequence, needing no neutral point and few components.

If V1 goes high, V2 being low, Tr1 remains cut off and Tr2 draws base current through R4. After 60°, phase 3 goes high and Tr2, already conducting, holds Tr1 off and led 2 lights during the 120° overlap to show the L1-L2-L3 sequence. In the reverse condition, V2 goes high while V1 is low and led 1 lights to show L1-L3-L2.

Cyril W W Palihawadana

Sana`a Republic of Yemen

Electronic fuse

Having a voltage range of 10-36V and handling currents up to 1A, this circuit disconnects a load in a time variable up to 100ms by changing a capacitor. Much greater currents and voltages can be handled by the same design with changed component values. It simply goes in series with load.

Most of the voltage drop across the circuit, V_{AB}, which is proportional to the DC load current and less than 2V, is across $R_{11,12}$. At switch on, all the supply voltage is across the fuse and Tr_3 conducts, its base current being supplied by R_4 and its collector current set by D_3 and R_8 according to $Ic_3 = (V_{D3}-V_{be3})/R_8$. Base current of Tr_4 is therefore stabilised, Tr_4 conducting and turning on Tr_5. Delay determined by C_1 prevents premature interruption of Tr_3 base current.

If load current increases excessively, the voltage dropped across R_{12} begins to turn Tr_2 on, reducing

309

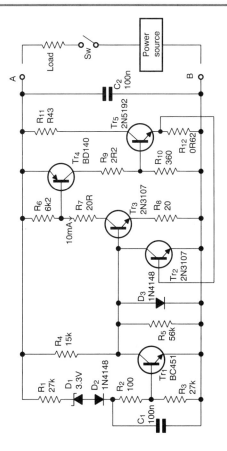

Figure 1. Circuit acting as a fuse for voltages from 10V to 36V and currents up to 1A, with values shown.

the collector current of $Tr_{3,4,5}$ and increasing the terminal voltage to more than 2V. When it exceeds 4.5V, D_1 avalanches, Tr_1 conducts and the cut-off of the three output transistors is cumulative, current through the fuse now being a few milliamps.

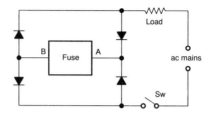

Figure 2. Same circuit operating in AC circuit.

Capacitor C_1 determines the time delay to cope with motor inrush currents or filament lamps and C_2 handles voltage spikes. Diode D_2 prevents C_1 discharging through the load when V_{AB} is almost zero.

With component changes, the circuit should be able to operate with currents from 10mA to 40A and on voltages from 6V to 500V. It can also be used as an AC fuse, as seen in Fig. 2.

To re-establish the circuit after an interruption, switch off for a short time.

N I Lavrentiev

Kaliningrad
Moscow Region Russia

Two-wire switch status detection

One central control determines the state of up to eight remote switches, using only two wires.

Figure 1 is the control unit, in which IC_1 is a *4094* latched shift register, driven by IC_2, a *4060* 14-stage binary counter/oscillator. Signals from IC_1 also drive the base of the power transistor Tr_1, which applies 12V to the signal bus at each positive excursion of the base drive.

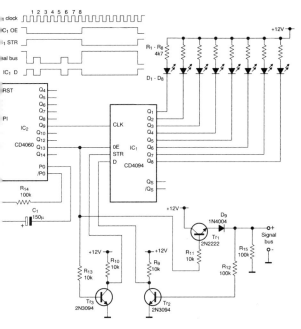

Figure 1. Control unit of remote switch state indicator for up to eight switches, using only two wires.

Remote units derive power from the bus, as shown in Fig. 2, and send pulses to the bus when the associated switch is off. When power is on the bus, C_3 charges to 5V through D_{10} and D_{12} and supplies power to the *4093 IC$_3$* when the bus is off. Capacitor C_2 also charges from the bus.

After eight clock periods, Tr_1 turns the bus off and C_2 discharges through $R_{16,17}$. IC_{3a} output goes high, this change being differentiated and passed to the bus as a square pulse whose width is set by the values of $C_4 R_{18}$ and after a delay determined by $R_{16,17}$. If the switch is on, no pulse passes IC_{3b}.

When pulses arrive on the bus at Tr_2, the *4094* D input goes low and the clock shifts the *4094* state. Delay time after bus power loss is set to a different period in each remote unit, so that the return pulse is detected at different clock times and the state of each remote switch is shifted in the *4094*. On the eigth pulse, the combined states are latched in the *4094* and illuminated leds indicate off switches.

Yongping Xia

Torrance
California
USA

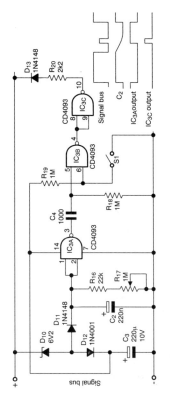

Figure 2. One of the remote switches (S1) and its associated pulse-forming circuitry. A return pulse passes to the signal bus when the switch is off.